T0106461

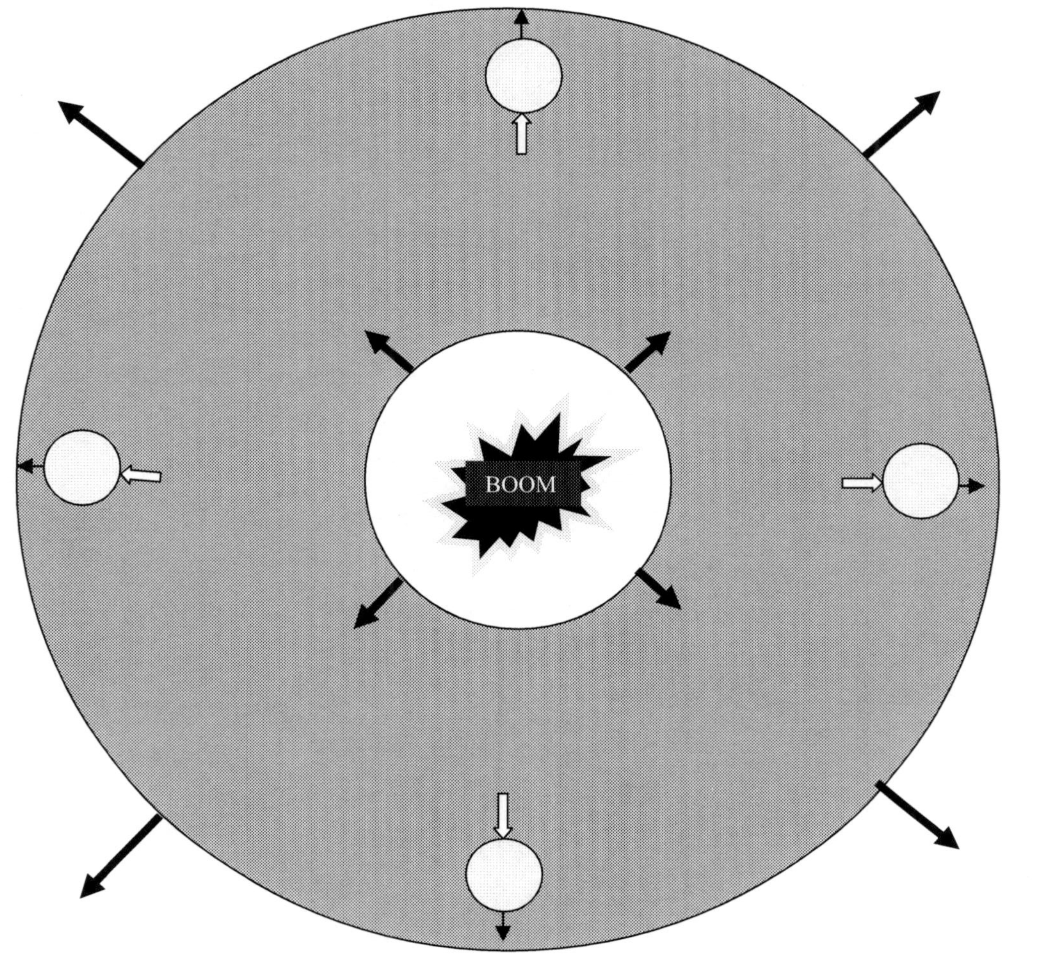

HYPERNOVA

PUSHING

SURROUNDING

STARS

HUBBLE SPACE TELESCOPE IDENTIFIES DARK ENERGY

A REVIEW OF ASTRONOMICAL OBSERVATORY DATA

AS IT RELATES TO DARK ENERGY

AND TECHNOLOGY TRANSFER SIMULATION OF DARK ENERGY

August A. Cenkner Jr.

B.A., B.S., M.S., Ph.D.

AuthorHouse™
1663 Liberty Drive
Bloomington, IN 47403
www.authorhouse.com
Phone: 1-800-839-8640

© 2009 August A. Cenkner Jr. All rights reserved.

No part of this book may be reproduced, stored in a retrieval system, or transmitted by any means without the written permission of the author.

First published by AuthorHouse 8/3/2009

ISBN: 978-1-4490-1134-5 (sc)

Library of Congress Control Number: 2009907731

Front Cover Background -- R. Williams and the Hubble Deep Field Team, STSci, NASA
Back Cover Background -- R. Williams and the Hubble Deep Field Team, STSci, NASA

Printed in the United States of America
Bloomington, Indiana

This book is printed on acid-free paper.

This book is a work of non-fiction. Unless otherwise noted, the author and the publisher make no explicit guarantee as to the accuracy of the information contained in this book and in some cases, names of people and places have been altered to protect their privacy.

To my wife Judy

She suggested that I write this book.
Her boundless energy and enthusiasm,
after all these years,
never ceases to amaze and inspire me.

TALK OUTLINE

Section	Topic	Page
1.	AUTHOR BACKGROUND	10
2.	HOW DARK ENERGY GOT ITS' NAME	11
3.	REVIEW TRAVELING-SHOCK-WAVE-PUSHING THEORY TO EXPLAIN DARK ENERGY	
	a. EXPLAIN THEORY	12
	b. REVIEW LABORATORY SIMULATIONS OF THEORY	13
	c. CONCLUSIONS FROM TRAVELING-SHOCK-WAVE-PUSHING SIMULATION	22
	d. OBSERVABLE EFFECTS OF TRAVELING-SHOCK-WAVE-PUSHING OF STARS	23
	e. WHAT HAPPENS NEXT – THE CAPTURE OF MOVING STARS	24
4.	THINGS TO LOOK FOR WHEN REVIEWING ASTRONOMICAL OBSERVATORY DATA	25
5.	REVIEW ASTRONOMICAL OBSERVATIONS THAT SUPPORT THEORY	26
	• HUBBLE OBSERVATION IDENTIFIES DARK ENERGY	35
	• FULL SCALE SPACE SIMULATION	36
6.	HOW DARK ENERGY AFFECTS THE HISTORY OF THE UNIVERSE	
	a. TIME LINE FROM THE BIG BANG OR RAPID EXPANSION	44
	b. THE FUTURE OF THE UNIVERSE	45
7.	HISTORY OF DARK ENERGY THEORY	47
8.	ABOUT THE AUTHOR	87

(1) AUTHOR BACKGROUND

August Cenkner Jr.

B.A. – Computer Science | B.S. – Aerospace Engineering
M.S. – Engineering Science | Ph.D. – Engineering Science

- Graduate School Major – Advanced Gas Dynamics
    - Supersonic Flow (1&2), Hypersonic Flow (1&2), Non-Steady Gas Dynamics, Rarefied Gas Dynamics, Molecular Flow, Radiation Gas Dynamics, Radiation Heat Transfer, Aerothermochemistry, Magnetohydrodynamics, Plasma Physics, Electromagnetic Theory, Electrodynamics, Direct Energy Conversion, Space Science, Diagnostic Techniques, Shock Tube Design and N.C.: Spectroscopy, Optics, Astronomy (1&2), Geology
- Professional Engineer / Adjunct Faculty UB School of Engineering / Scientist / Amateur Astronomer
    - Courses Taught (32): Statics, Dynamics, Thermodynamics, Fluid Mechanics, Vibrations, Heat Transfer
- Interdisciplinary Background
    - Gas Dynamics / Physics / Astronomy / Computers
- 40 Years Research and Development Experience
    - Theoretical / Experimental / Design Work
    - 21 Technical Publications in Open Literature
    - Strong Gas Dynamics Background (Theoretical and experimental)
- Currently Conducting Independent Unfunded Research on Dark Energy
    - Working More Than Five Years
    - Applying My Gas Dynamic Training and Experience to Astronomical Problems (***TECHNOLOGY TRANSFER***)
    - Spent ~ $10,000. Of My Own Money
    - Created Original Classical Dark Energy Theory That Explains Currently Unexplained Astronomical Phenomena

Dark Energy Project Activities (Section 7)

(1) Cenkner, August A. Jr., "Dark Energy Identified", Sky and Telescope, 7/8/08, pg. 85.
(2) Cenkner, August A. Jr., "Dark Energy – Laboratory Simulations Lead to Predictions of: Star Accelerations; Creation of Spiral Galaxies; Formation of Voids, Walls, and Clusters", 2nd edition, ISBN 978-1-4343-0661-6 (sc), 194 pgs, AuthorHouse, 8/22/07.
(3) Cenkner, August A. Jr., "A Dark Energy Theory", Origins of Dark Energy Conference, McMaster University, Hamilton, Ontario, Canada, 6/14/07.
(4) Cenkner, August A. Jr., "A Dark Energy Theory Correlated With Laboratory Simulations and Astronomical Observations", ISBN 1-4208-3447-9 (sc), 98 pgs,. AuthorHouse, 8/23/05.
(5) Cenkner, August A. Jr., "Galaxies", Buffalo Astronomical Association, Buffalo State College, Buffalo, NY, 6/12/05.
(6) Cenkner, August A. Jr., "Universe Acceleration Predictions Using New Dark Energy Explanation", Buffalo Astronomical Association, Buffalo NY, 7/10/04.
(7) Cenkner, August A. Jr., "A Theory to Explain Dark Energy", The Spectrum, Buffalo Astronomical Association, Buffalo State College, Buffalo NY, 3/7/04.
(8) Cenkner, August A. Jr., "Dark Energy", Sky and Telescope, 2/08, pg.80.

(2) HOW DARK ENERGY GOT ITS' NAME

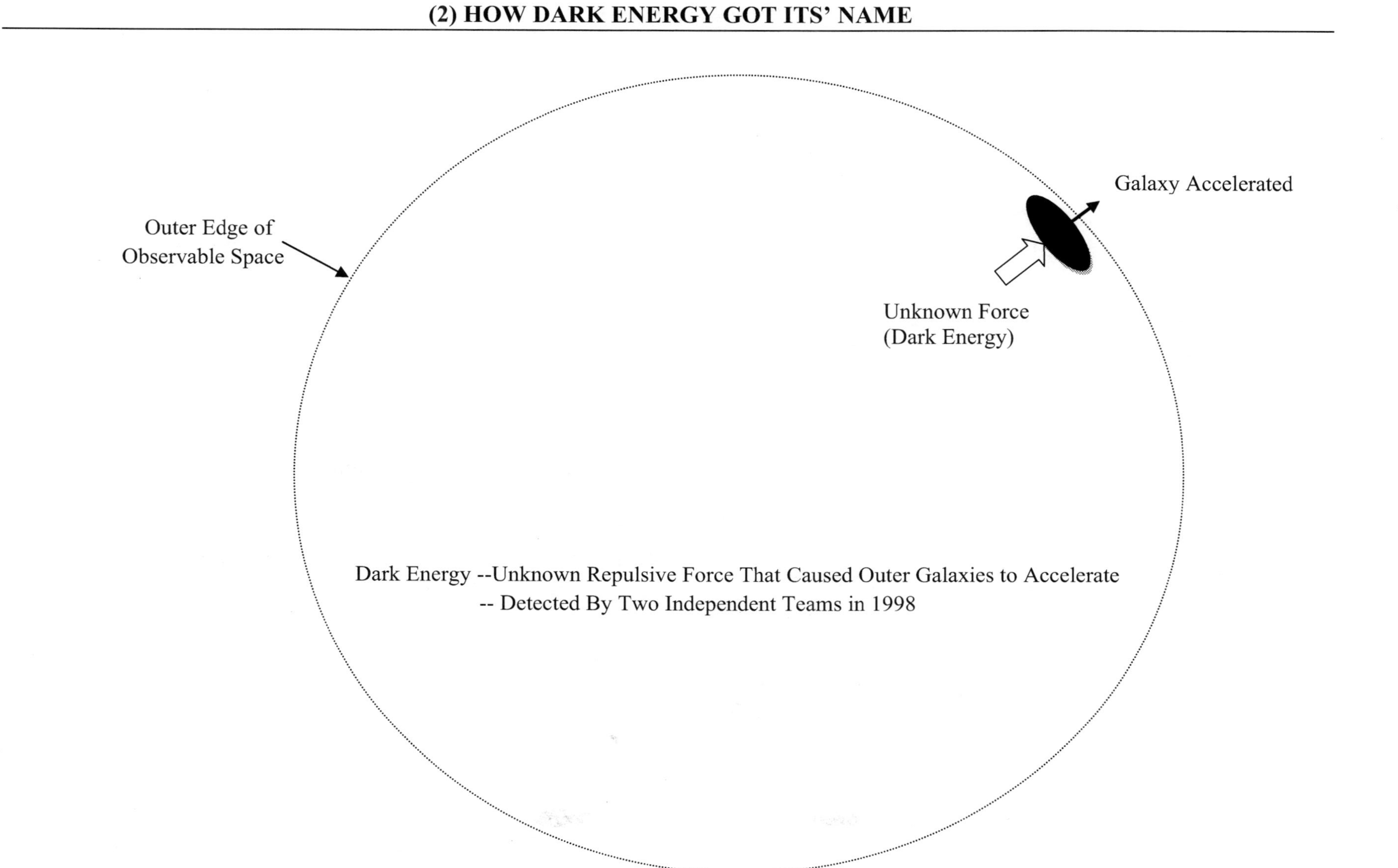

(3) TRAVELING-SHOCK-WAVE-PUSHING THEORY FOR DARK ENERGY

a. Explain Theory

- Traveling-Shock-Wave-Pushing Theory For Dark Energy
  - Dark energy is the energy contained in traveling-gas-cloud-shock-waves
    - Upon impact with star, shock wave energy is transferred to star kinetic energy
    - This occurs when gas cloud pressure pushes star, due to creation of low pressure wake
  - Gas-cloud-shock-wave is caused by violent gas ejecting processes like star explosions, star collisions, etc.
  - Accelerated stars are captured, into orbit, by other stars and entities

b. Review Laboratory Simulations of Theory

Consider a Star Explosion That Creates a Traveling-Spherical-Shock-Wave (Gas Cloud)

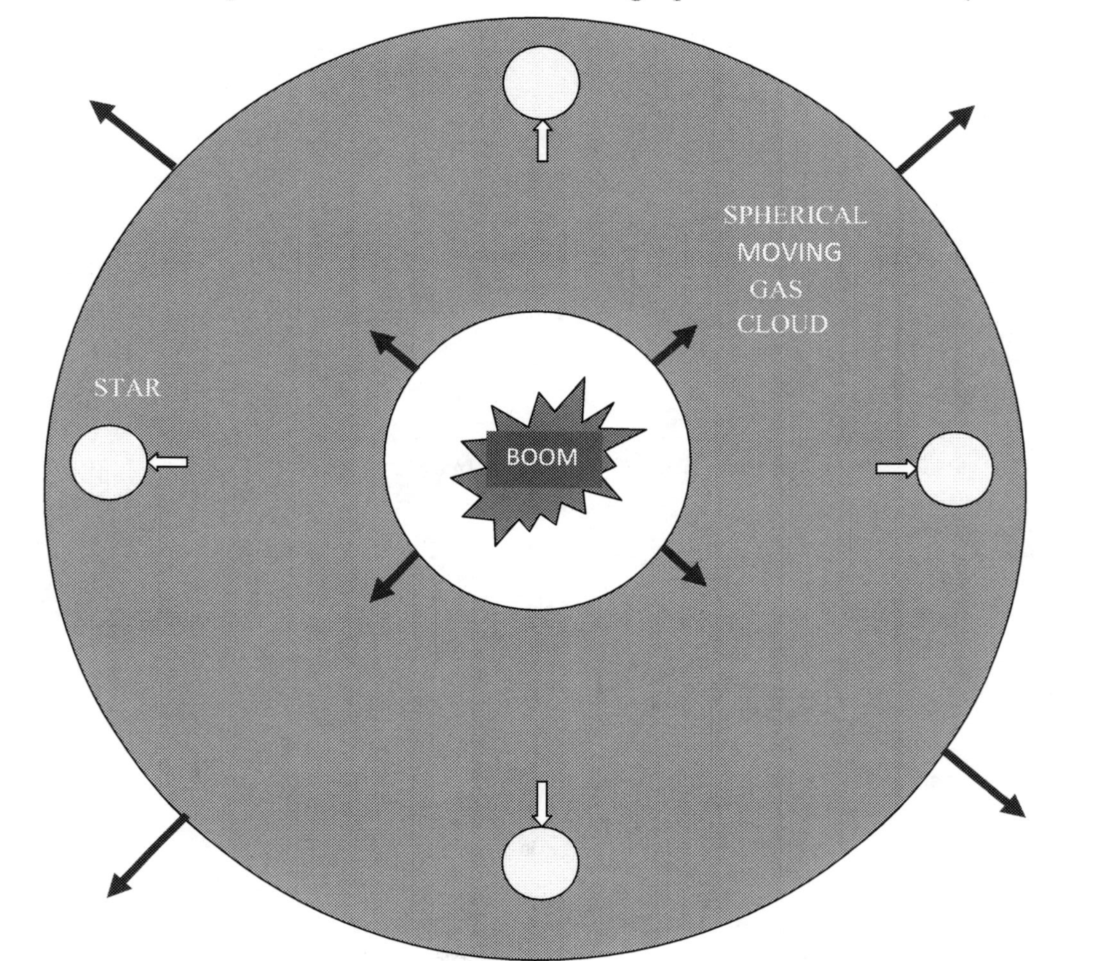

Key Question: What Happens When Traveling-Gas-Cloud-Shock-Wave Impacts Stars?

(3) TRAVELING-SHOCK-WAVE-PUSHING THEORY FOR DARK ENERGY

b. Review Laboratory Simulations of Theory

What Happens When Traveling-Shock-Wave (Gas Cloud) Impacts a Star?

Employ Simulations to Find Out

Section	Simulation	Page
3b-1	Bench Top Simulation (Cenkner)	15
3b-2	Subsonic Plasma Wind Tunnel Simulation (Cenkner) $M < 1$	16
3b-3	Supersonic Wind Tunnel Simulation (Cenkner) $M > 1$	19
3b-4	Hypersonic Computer Model (Gordon, et. al.) $M >> 1$	20
3b-5	Atomic Bomb Explosion	21
5-2	Full Scale Space Simulation by Hubble	36

b. Review Laboratory Simulations of Theory

Simulation No. 1 – Bench Top Simulation Using Fan

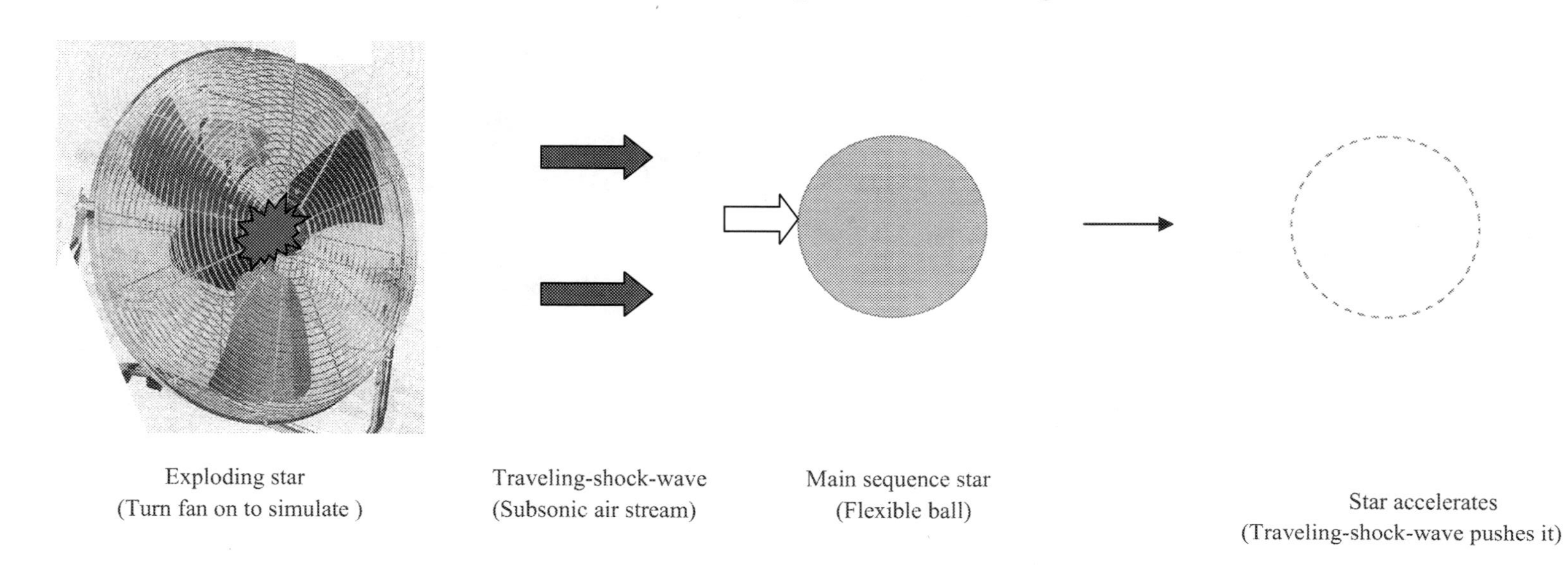

<u>Conclusion</u>

When a fan is turned on, the resulting subsonic traveling-shock-wave will push a ball, in its' path, upon impact.

or

An exploding star, that ejects gas, will push another star when the traveling-gas-cloud-shock-wave impacts it.

or

*Internal energy ⇨ Kinetic energy of traveling gas cloud ⇨ Kinetic energy of star

* Ref. 2 uses the First Law of Thermodynamics for a more sophisticated energy analysis.

(2) TRAVELING-SHOCK-WAVE-PUSHING THEORY FOR DARK ENERGY

b. Review Laboratory Simulations of Theory

Simulation No. 2 -- Subsonic Plasma Wind Tunnel (Cenkner)*. M<1

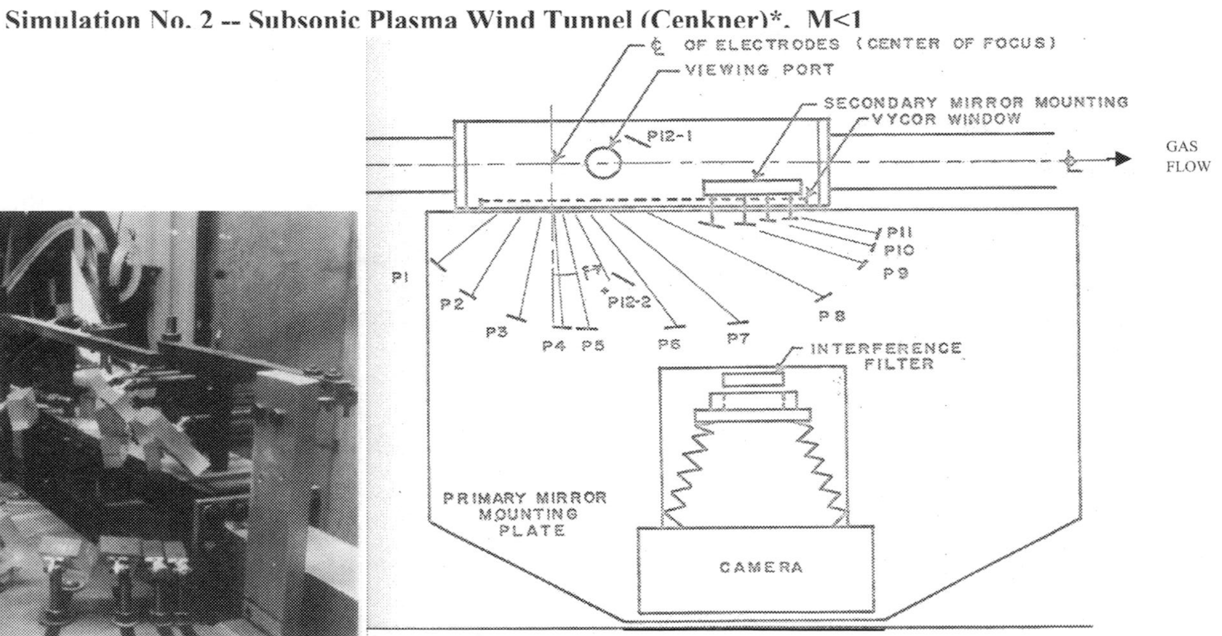

UNIQUE OPTICAL SYSTEM

* Ref. 2 & 4

b. Review Laboratory Simulations of Theory

Simulation No. 2 -- Subsonic Plasma Wind Tunnel (Cenkner), M<1

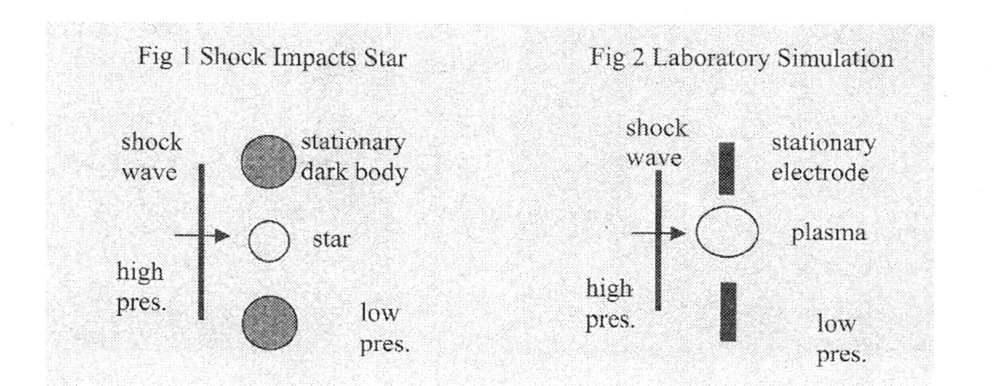

Star constrained by gravitational force

Plasma (gas) constrained by electric field

b. Review Laboratory Simulations of Theory

Simulation No. 2 -- Subsonic Plasma Wind Tunnel (Cenkner)

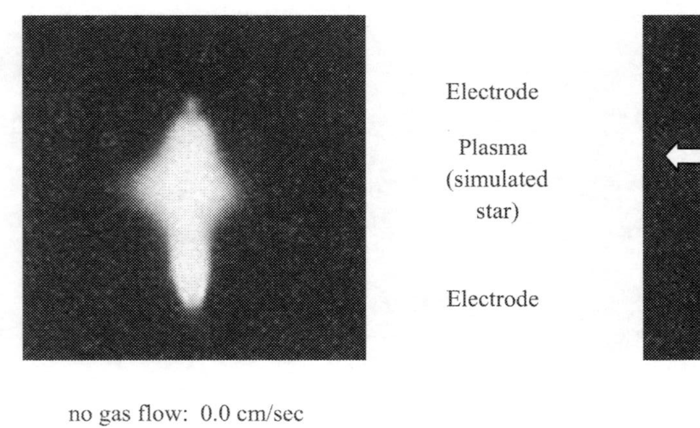

no gas flow: 0.0 cm/sec

gas flow: 127 cm/sec

Conclusion

In a subsonic gas stream, a main sequence star will act like a solid deformable body and a low pressure wake will appear on the downstream side. This will cause the star to move, because of an unbalanced force on the star.

b. Review Laboratory Simulations of Theory

Simulation No. 3 -- Supersonic Wind Tunnel (Cenkner)*, M>1

Conclusion

*Ref 4

In a supersonic gas stream, a gaseous sphere will act like a solid deformable body and a low pressure wake will appear on the downstream side.

b. Review Laboratory Simulations of Theory

Simulation No. 4 -- Computer Model of Hypersonic Gas Flow Past Solid Sphere (Van Dyke, et. al.)*, M>>1

Pressure Ratio Around Sphere

$\left\{ \frac{\text{Surface pressure}}{\text{Stagnation Pressure}} \right\}$

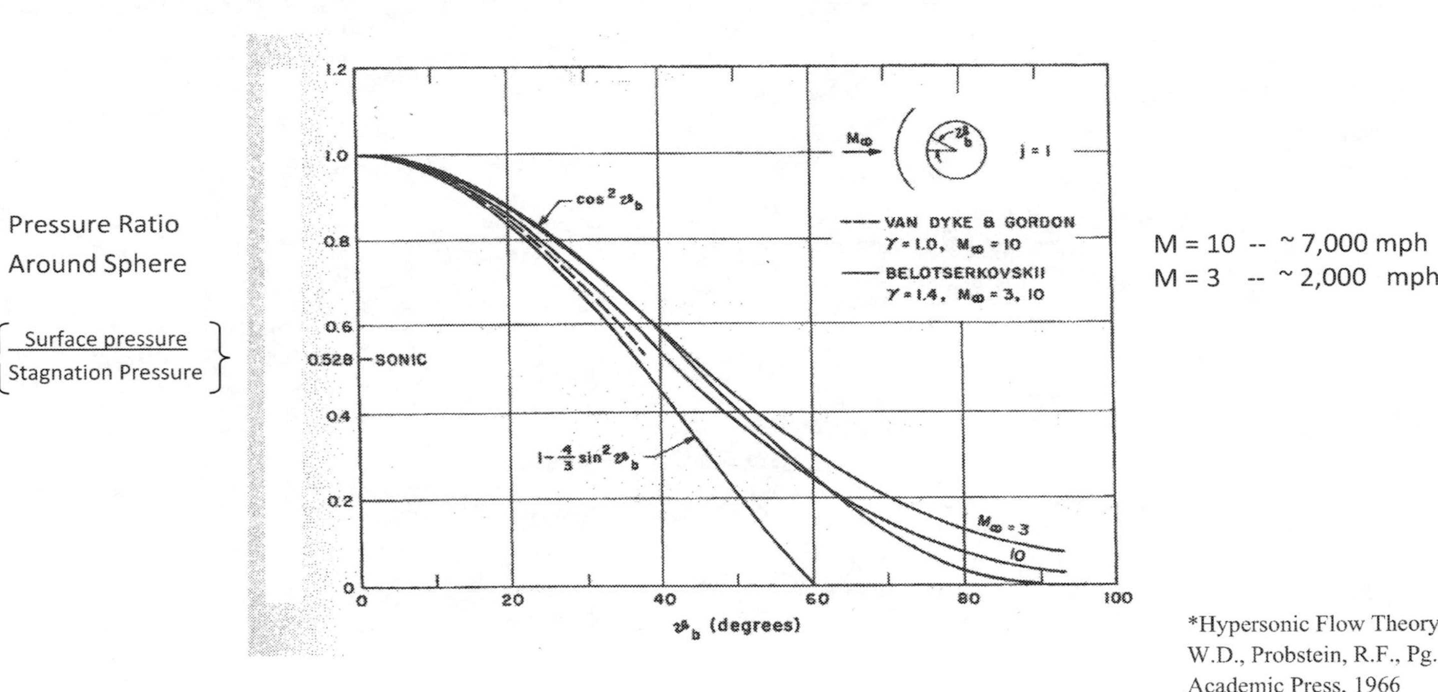

M = 10 -- ~ 7,000 mph
M = 3 -- ~ 2,000 mph

*Hypersonic Flow Theory, Hayes, W.D., Probstein, R.F., Pg. 423 Academic Press, 1966

Conclusion

In a supersonic and hypersonic gas stream, a low pressure wake is formed on the downstream side of a solid sphere, producing a force unbalance.

(3) TRAVELING-SHOCK-WAVE-PUSHING THEORY FOR DARK ENERGY

b. Review Laboratory Simulations of Theory

Simulation No. 5 -- *Atomic Bomb Explosion

*static.desktopnexus.com

Conclusion

A high speed traveling-gas-cloud-shock-wave, created by a powerful explosion, pushes everything in its path.

(3) TRAVELING-SHOCK-WAVE-PUSHING THEORY FOR DARK ENERGY

c. Conclusions from Traveling-Shock-Wave-Pushing (Dark Energy) Simulations

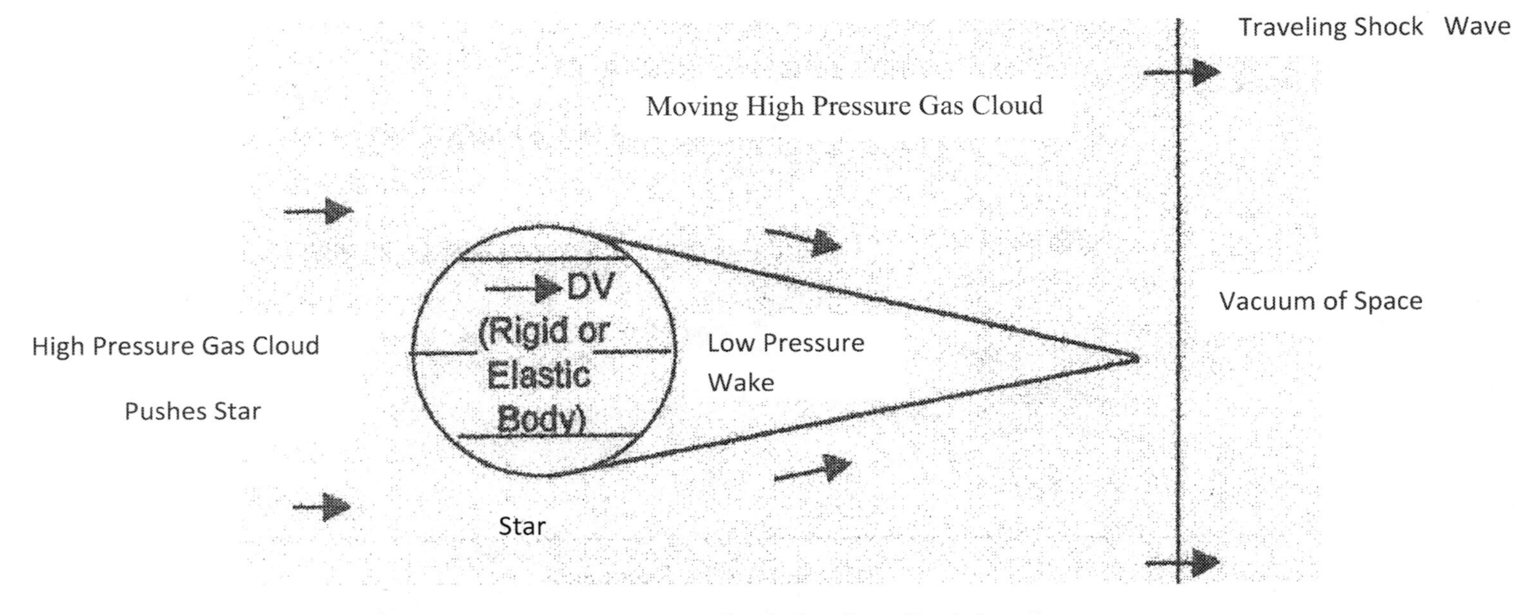

Conclusions From Simulations

- *Star behaves like ***solid deformable elastic body***
- A ***low pressure wake*** is formed downstream of the star, in subsonic, supersonic, and hypersonic flow
- This pressure differential causes the star to be pushed and, hence, accelerated
- Since pressure is a classical force, can apply Newton's Law's: $F = m A$ or I = Impulse $= \underline{F} \times (\Delta t) = m \times (\Delta V)$

Note: (1) ΔV increases as $\underline{F}$ or Δt increases

(2) ΔV same for same $\underline{F} \times (\Delta t)$

(3) Same I for different combinations of m and ΔV

* This conclusion supported by Hubble discovery of stars, moving through a gas cloud at hypersonic speed , that have an oblique bow shock wave; see Section 5: Evidence 5-2.

d. Observable Effects, From Earth, of Traveling-Shock-Wave-Pushing of Stars

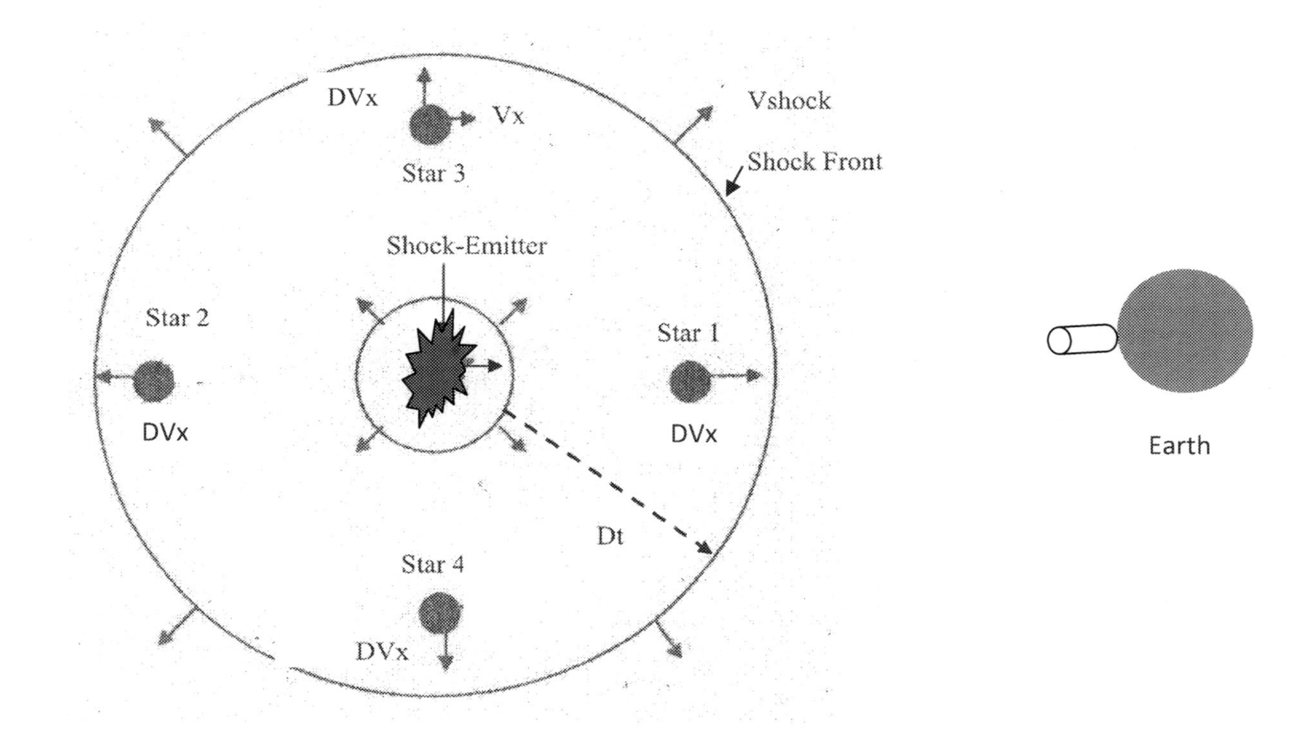

1. All stars in spherical volume will accelerate in radial direction
2. All stars in spherical volume will be removed from volume
3. Some stars will move toward earth and some away from earth (nos. 1 & 2) → ←
4. All stars move at constant velocity, along straight line, after shock passes
5. Some stars will move across the sky (nos. 3 & 4) ↓ ↑
6. Many of these accelerated stars will be captured by other stars or bodies
7. Shock emitter residue (dark matter ???) will remain at center of sphere; may be visible or invisible

e. *What Happens Next -- The Capture of Moving Stars

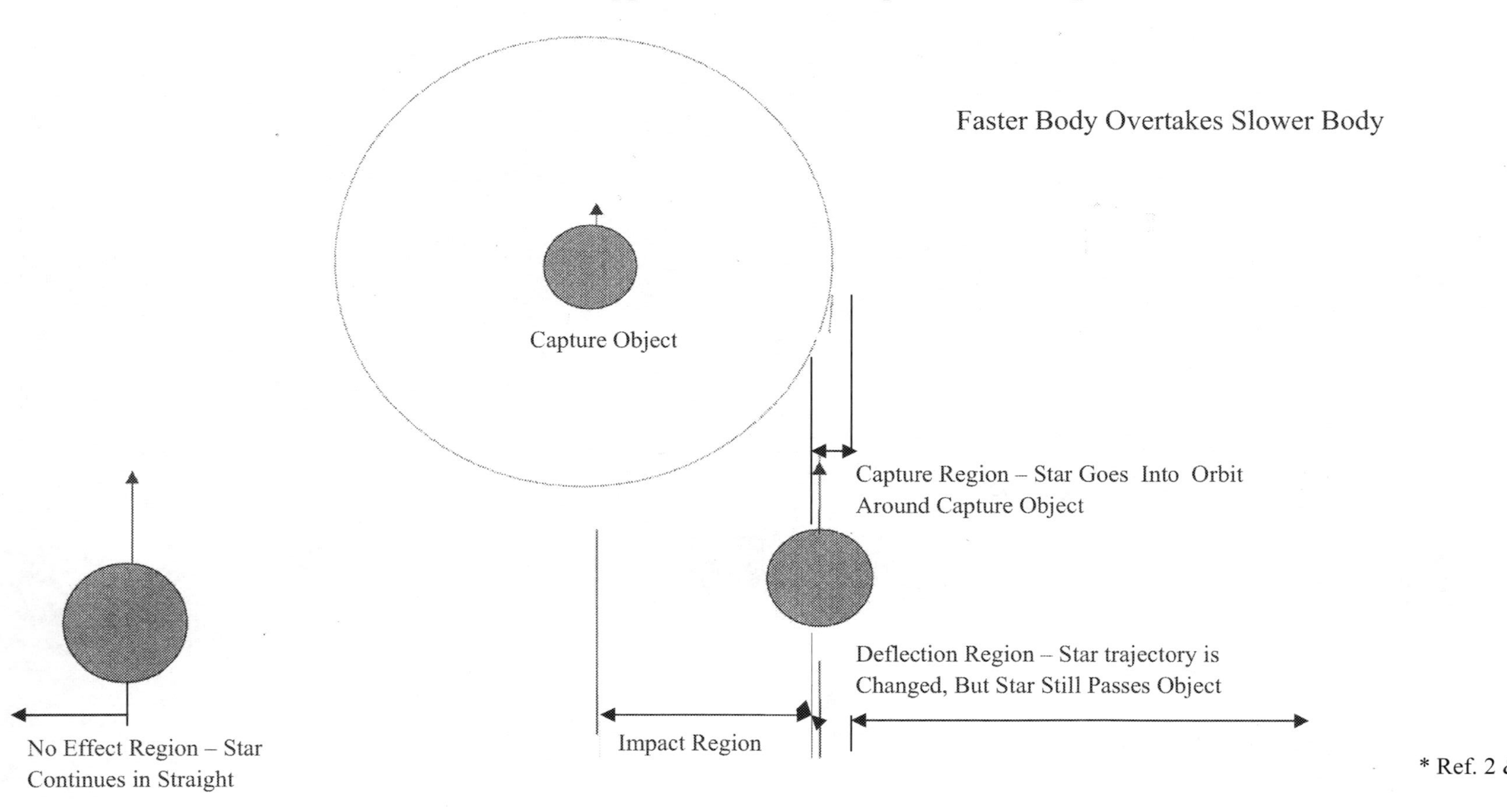

* Ref. 2 & 4

Faster Stars (or star groups) Captured By Slower Stars, or Bodies, to Form Revolving Star Groups

- Binary, Tertiary, Multiple Revolving Star Groups
- Open Clusters, Globular Clusters, Galactic Clusters, Halo, etc.
- Galaxies

(4) THINGS TO LOOK FOR WHEN REVIEWING ASTRONOMICAL OBSERVATORY DATA

1. Star Acceleration 26
   - Acceleration of Early Galaxy
   - Tycho's Supernova
2. Volume of Space Cleared of Stars 27
   - Voids
   - Hoag's Type Ring Galaxies
3. Stars or Star Groups Revolving Around Other Objects 29
   - Binary Stars, Open Clusters, Globular Clusters, Galaxy Core, All Galaxies, Milky Way Halo
4. Star or Galaxy Collisions 31
   - Collision of Spiral Galaxies in Wall
   - Collisions of Numerous Galaxies
5. Unusual Star Motions 34
   - Galaxy Streams -- Groups of Galaxies Moving Across The Sky
   - Stars Speeding Across the Sky
     - Hubble Identifies Dark Energy (Section 5-2)
6. Stars or Star Groups Moving Toward or Away From Earth 37
   - Motion of Galaxies – Hubble's Law
7. Complicated Star Motions Difficult to Explain 38
   - Spiral Galaxies
   - Compact Galaxies in Early Universe

(5) REVIEW ASTRONOMICAL OBSERVATIONS THAT SUPPORT TRAVELING-SHOCK-WAVE-PUSHING THEORY

Evidence Supporting Item 1 --- Star Acceleration

1. 1998 detection of acceleration of galaxies in outer limits of visible universe (see Section 2)
2. Tycho's Supernova (SN1572)
   - Binary star system
   - Observed in 1572 (and probably before that)
   - Surviving star (Tycho G)
     - ***Traveling at more than 40 times faster than other neighbors***
   - Shock wave still visible (20 light years diameter)

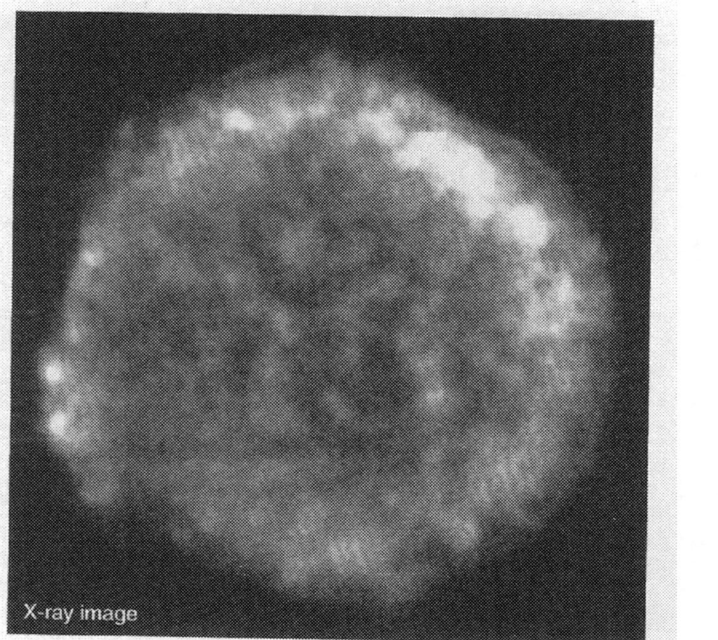

*Spherical Remnant From Supernova (expanding in radial direction)

Pushing-Gas-Cloud –Shock-Wave Theory

- Supernova creates spherical gas cloud shock wave
- Shock wave pushes binary companion star in radial direction, causing it to accelerate
- Different size stars accelerate to different velocities
- The spherical remnant from the supernova is actually the dark energy remnant

*M.A. Seeds, Horizons: Exploring the Universe, Brooks/Cole –Thompson Learning, 2004, pg. 220.

(5) REVIEW ASTRONOMICAL OBSERVATIONS THAT SUPPORT TRAVELING-SHOCK-WAVE-PUSHING THEORY

Evidence Supporting Item 2 --- Volume of Space Cleared of Stars

Example I – Voids

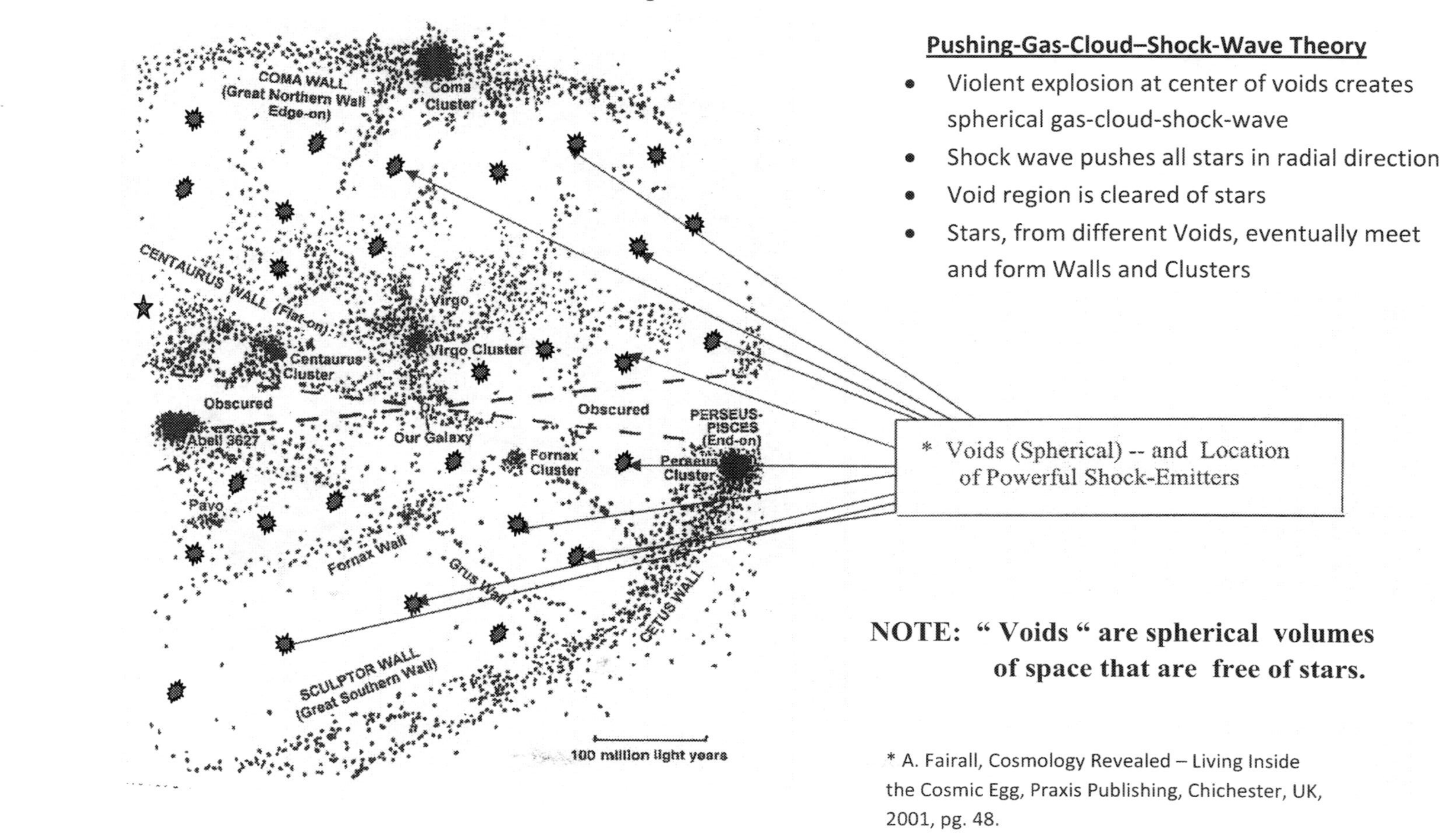

Pushing-Gas-Cloud–Shock-Wave Theory

- Violent explosion at center of voids creates spherical gas-cloud-shock-wave
- Shock wave pushes all stars in radial direction
- Void region is cleared of stars
- Stars, from different Voids, eventually meet and form Walls and Clusters

NOTE: " Voids " are spherical volumes of space that are free of stars.

* A. Fairall, Cosmology Revealed – Living Inside the Cosmic Egg, Praxis Publishing, Chichester, UK, 2001, pg. 48.

(5) REVIEW ASTRONOMICAL OBSERVATIONS THAT SUPPORT TRAVELING-SHOCK-WAVE-PUSHING THEORY

Evidence Supporting Item 2 --- Volume of Space Cleared of Stars

Example II –*Hoag's Type Ring Galaxies

Stars Moving Outward In Radial Direction

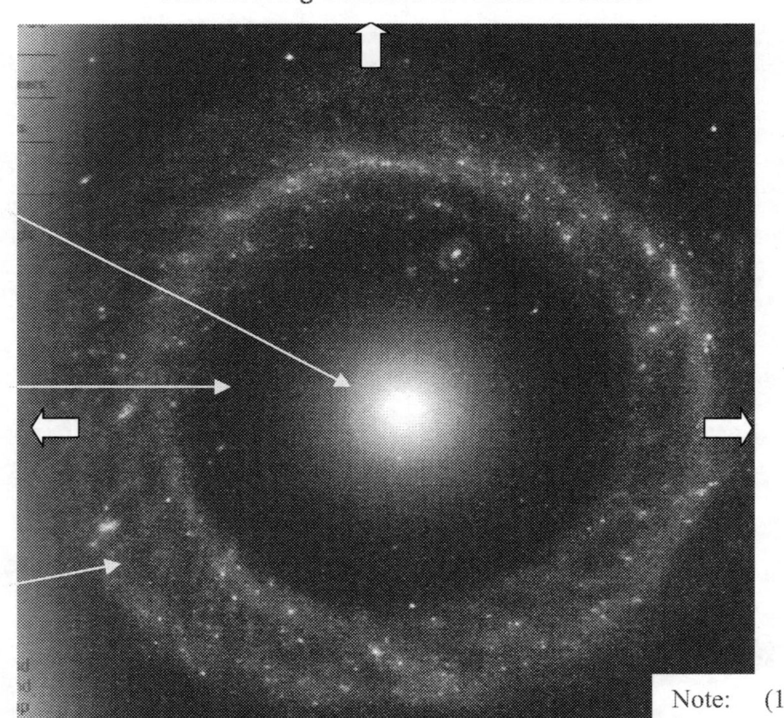

Ball of Older

Inner Ring Without Stars

Outer Ring of Stars

No Galaxies Close Enough For Past Collision

Pushing-Gas-Shock-Wave Formation Theory

1. Star explodes at center of disk galaxy, creating spherical shock wave.
2. Shock wave pushes stars outward radially, clearing inner volume of stars.
3. Remnant of star explosion (dark body) captures older stars to form ball of older stars

Note:

(1) Some ring galaxies have core in central region:
- Black hole
- Smaller galaxy

(2) Some ring galaxies thought to be product of galaxy collisions, with a second galaxy passing through center of disk galaxy.

(3) Have second ring galaxy in background.

*Rees, M., Universe: The Definitive Visual Guide, DK Publishing Inc., pg. 309

(5) REVIEW ASTRONOMICAL OBSERVATIONS THAT SUPPORT TRAVELING-SHOCK-WAVE-PUSHING THEORY

Evidence Supporting Item 3 --- Stars or Star Groups Revolving Around Other Objects

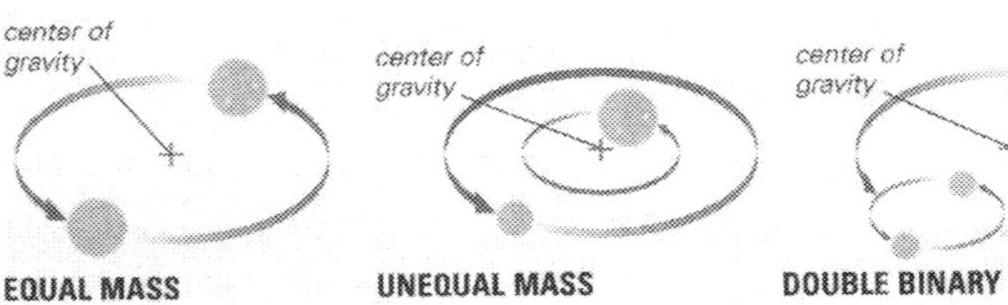

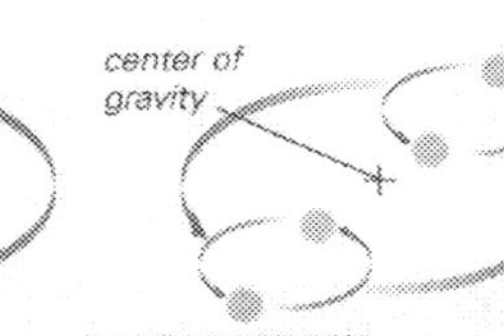

Revolving Groups of Stars

Binary Star Group Revolving Around Center of Mass

Galaxy Cluster

Group of Galaxies Revolving Around Center of Mass

(from Kitt Observatory)

Revolving Groups of Stars

- Open Clusters
- Globular Clusters
- Core of Milky Way Galaxy
- Halo Around Milky Way Galaxy

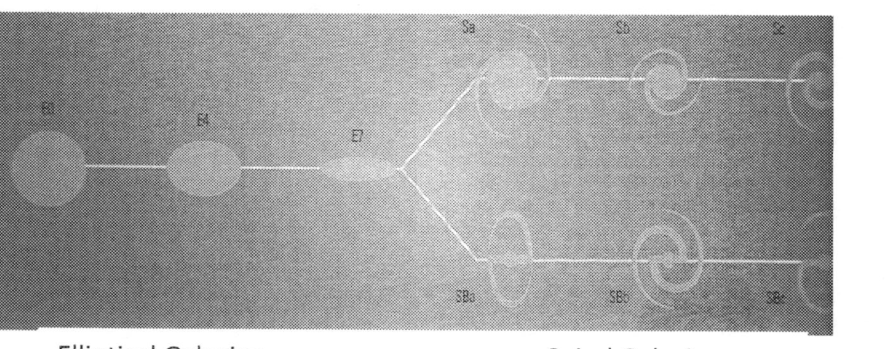

Elliptical Galaxies Spiral Galaxies

(Astronomy Encyclopedia, Moore)

Revolving Groups of Stars

Numerous Types of Galaxies

(Elliptical, Spiral, Irregular, Disc, Ring)

*M.A. Seeds, Horizons: Exploring the Universe, Brooks/Cole –Thompson Learning, 2004, pg.

(5) REVIEW ASTRONOMICAL OBSERVATIONS THAT SUPPORT TRAVELING-SHOCK-WAVE-PUSHING THEORY

Evidence Supporting Item 3 --- Stars or Star Groups Revolving Around Other Objects

Estimated Size and Population of Galaxies and Star Groups

2004 Estimated Size and Population of Galaxies*

Galaxy type	No. of stars	Galaxy diameter (ly)	Comment
Spiral			Contain young and old stars
Most spiral		<< D_{MW}	Much smaller than Milky Way
Milky Way	> 200 Billion	D_{MW} = 100,000	Large compared to most spirals
Large spiral		4 x D_{MW}	A few are larger than Milky Way
Irregular		(1-25%) D_{MW}	Contain young and old stars. Large clouds of gas and dust.
Elliptical			Little gas and dust. No young stars.
Giant		5 x D_{MW}	
Dwarf		1% D_{MW}	

* ~ 100 billion galaxies in observable universe; ~70% are spirals; a few per cent are irregulars; the rest are ellipticals

Estimated Size and Population of Star Clusters*

Star cluster type	No. of stars	Cluster diameter (ly)	Comment
Open cluster	10 - 1000	81.5	Stars in cluster have same age but different mass
Globular cluster	10^5 - 10^6	32.6 – 97.8	

*May Be Outdated

*M.A. Seeds, Horizons: Exploring the Universe, Brooks/Cole –Thompson Learning, 2004, pg. 235

Evidence Supporting Item 4 --- Galaxy or Star Collisions

Evidence 4-1 -- Cause of Galaxy Collisions in Walls

Review Section of Sponge-Like Structure of Universe

Would expect to find galaxies revolving around other galaxies

Shock wave pressure, from explosions in Voids, pushing stars/galaxies

Different galaxies, from adjacent voids, being pushed in opposite direction

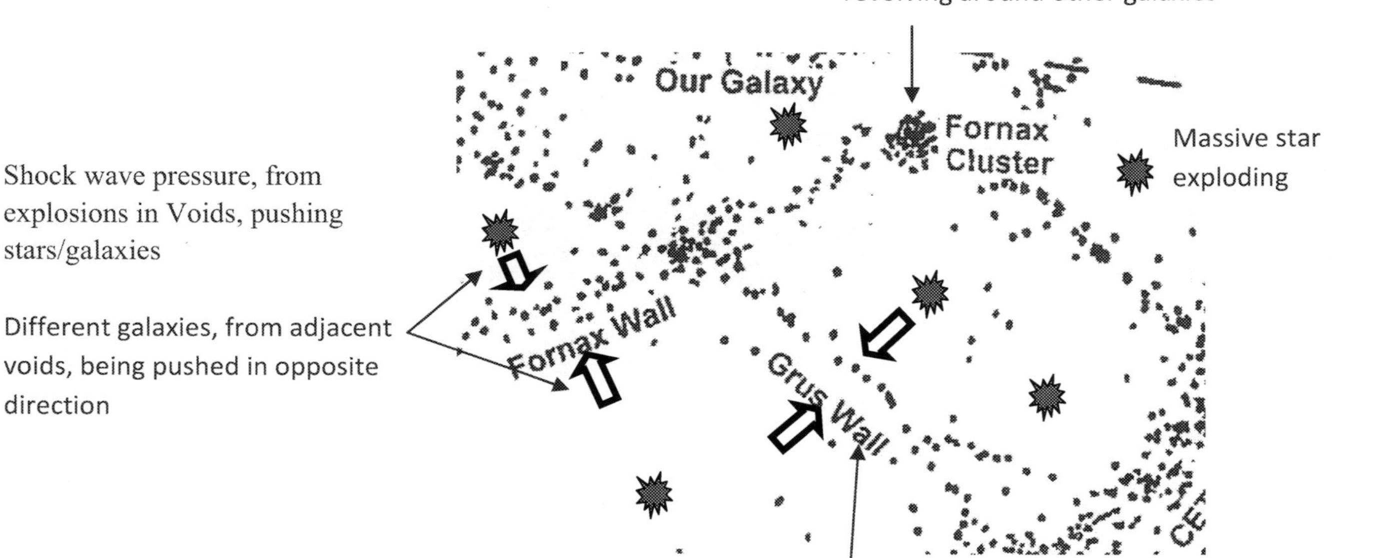

Massive star exploding

Would expect to find colliding and revolving galaxies in walls and galaxy groups, with galaxies originating from different Voids

Evidence Supporting Item 4 --- Galaxy or Star Collisions

Evidence 4-1 -- Collision of NGC6050 and IC 1179 Spiral Galaxies in Wall

Hubble Image: heio 08100p

- Part of Hercules Galaxy Cluster
- Which is part of Great Wall of Clusters and Superclusters (largest known structure in the universe)
- Located in constellation of Hercules
- 450 million light years from earth

Evidence Supporting Item 4 --- Galaxy or Star Collision

Evidence 4-2 -- Collisions/Interactions of Numerous Galaxies

Hubble
Images of colliding galaxies

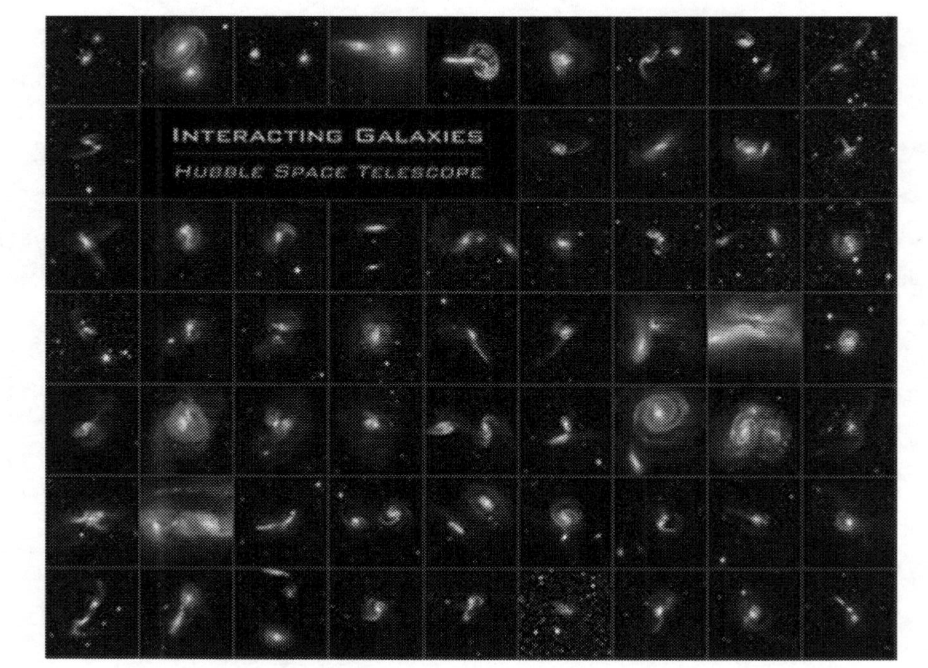

<u>Forces Driving Galaxies to Collide</u>

(1) Attraction, due to gravitation
  a. Between galaxies
  b. With some outside bodies
(2) Pushing, due to traveling-shock-wave-pressure
(3) Combination of both

(5) REVIEW ASTRONOMICAL OBSERVATIONS THAT SUPPORT TRAVELING-SHOCK-WAVE-PUSHING THEORY

Evidence Supporting Item 5 --- Unusual Star Motions

Evidence 5-1 *Galaxy Streams – Groups of Galaxies Moving Across the Sky

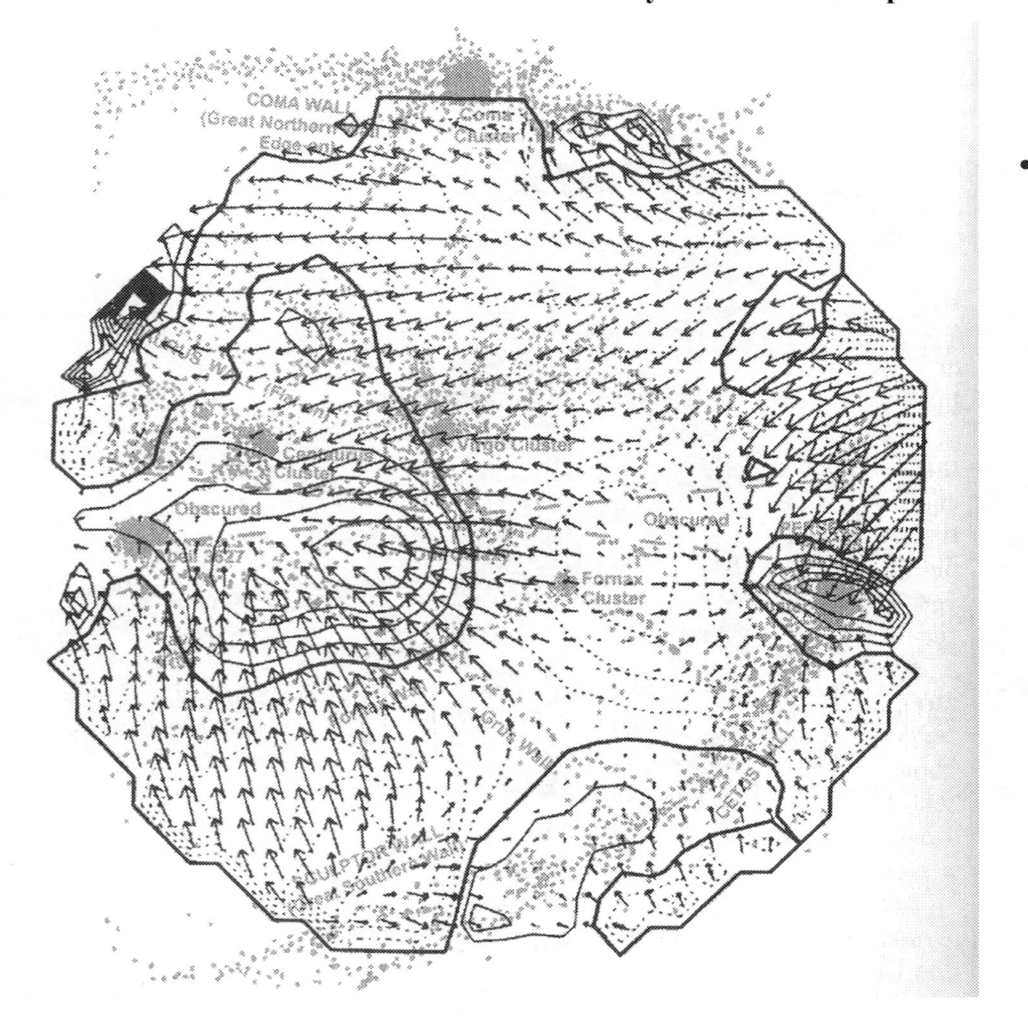

Arrows

- Show the movement of galaxies relative to Walls and Superclusters*
  - Galaxies appear to originate from Voids
  - Galaxies appear to be moving toward Walls

Galaxy Stream motion

- Supports theory that the galaxies were pushed from Voids by traveling-shock-wave-pushing
- Pushing is the result of explosions that created traveling-shock-wave-pushing-pressure

*A. Fairall, Cosmology Revealed – Living Inside the Cosmic Egg, Praxis Publishing, Chichester, UK, 2001, pg. 80.

(5) REVIEW ASTRONOMICAL OBSERVATIONS THAT SUPPORT TRAVELING-SHOCK-WAVE-PUSHING THEORY

Evidence Supporting Item 5 --- Unusual Star Motions

Evidence 5-2 Stars Speeding Across The Sky – Identify Dark Energy

Hubble

- One of 14 high speed stars caught streaking across sky through region of dense gas
- Bow shock could be 100 billion to a trillion miles wide
- Traveling at more than 112,000 miles per hour

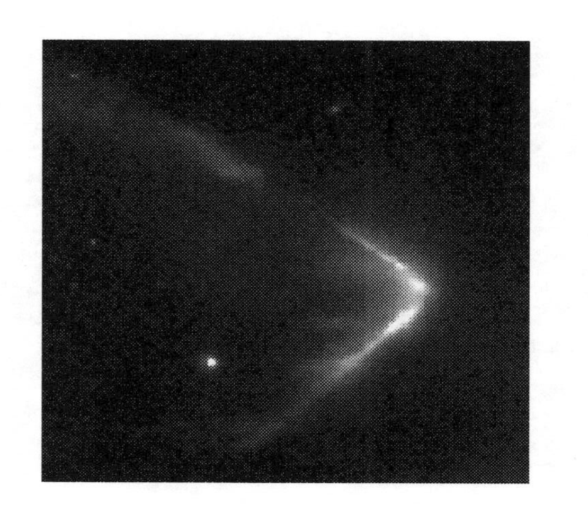

Hubble: STScI-PRC2009-03a

Traveling-Shock-Wave-Pushing Theory (see next page)

- Star is moving from left to right, across the sky
- It may be part of a galaxy stream
- Traveling-shock-wave pushing, at some earlier time, can account for this high velocity, the oblique shock wave, and the unusual movement across the sky

HUBBLE IDENTIFIES DARK ENERGY

- The ***star behaves like a solid body*** (which reinforces conclusions from laboratory simulations, summarized in Section 3c)
- A detached oblique shock is characteristic of a solid spherical object that is moving at hypersonic speed through a gas cloud; ***see next page for laboratory simulation***
- A ***low pressure wake*** will therefore exist when a traveling shock wave impacts a main sequence star. Due to the resulting pressure differential, the star will be pushed

Evidence Supporting Item 5 --- Unusual Star Motions

Evidence 5-2 Stars Speeding Across The Sky

*Shadowgraph Image of ½ in. Solid Sphere Moving at Hypersonic Speed Through Air

(For comparison with star image on previous page. Shows that ***star acts like a solid sphere, with a low pressure wake***)

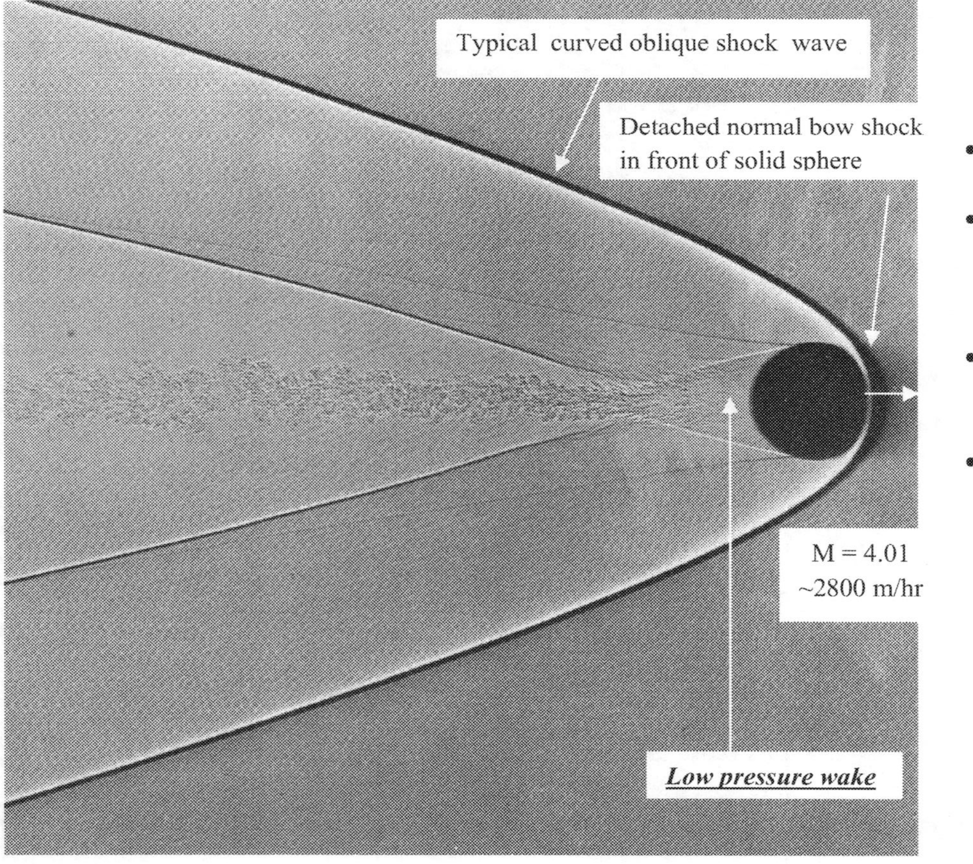

HUBBLE IDENTIFIES DARK ENERGY

- Here, we see the flow field around a solid sphere that is moving through air at hypersonic speed, from left to right.
- The flow field would be identical if we were looking at the flow around a stationary sphere that has hypersonic gas flowing past it, from right to left. (Hence the use of wind tunnels to test small scale models of aircraft and missiles)
- Therefore, Hubble image: StSci-PRC2009-03a, on the previous page, is ***actually a full scale space simulation of the traveling-shock-wave-pushing theory*** that is proposed for dark energy.
- Hubble image: StSci-PRC2009-03a, previous page, in effect shows that a ***low pressure wake will be formed downstream*** of a main sequence star when it is impacted by a traveling shock wave. This reinforces the conclusions that were drawn from the simulations introduced in Section 3.

*M. Van Dyke, An Album of Fluid Motion, Parabolic Press, 1982, Pg 166.

(5) REVIEW ASTRONOMICAL OBSERVATIONS THAT SUPPORT TRAVELING-SHOCK-WAVE-PUSHING THEORY

Evidence Supporting Item 6 --- Stars or Star Groups Moving Toward or Away From Earth

Recessional Motion of Galaxies – Hubble's Law

* Radial Peculiar Velocity = Radial Actual Velocity - Radial Recessional Velocity

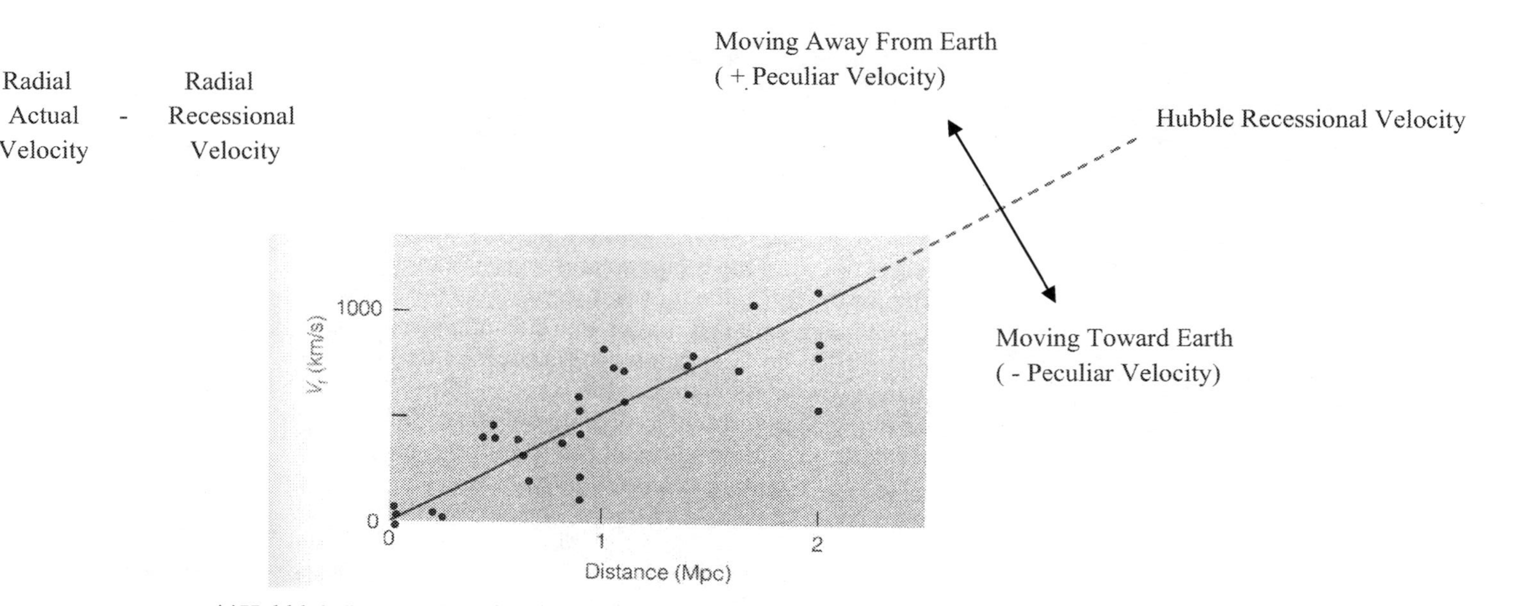

**Hubble's Law – Recessional Velocity of Galaxy Vs Distance From Earth

*Radial Peculiar Velocity = Radial velocity of galaxy caused by earlier pushing of galaxy by traveling-shock-wave; can move toward or away from earth.

**M.A. Seeds, Horizons: Exploring the Universe, Brooks/Cole –Thompson Learning, 2004, pg. 263.

(5) REVIEW ASTRONOMICAL OBSERVATIONS THAT SUPPORT TRAVELING-SHOCK-WAVE-PUSHING THEORY

Evidence Supporting Item 7 --- Complicated Star Motions Difficult to Explain

Evidence 7-1 --- Spiral Galaxies – The Milky Way Galaxy

*Artist Conception
Milky Way Galaxy

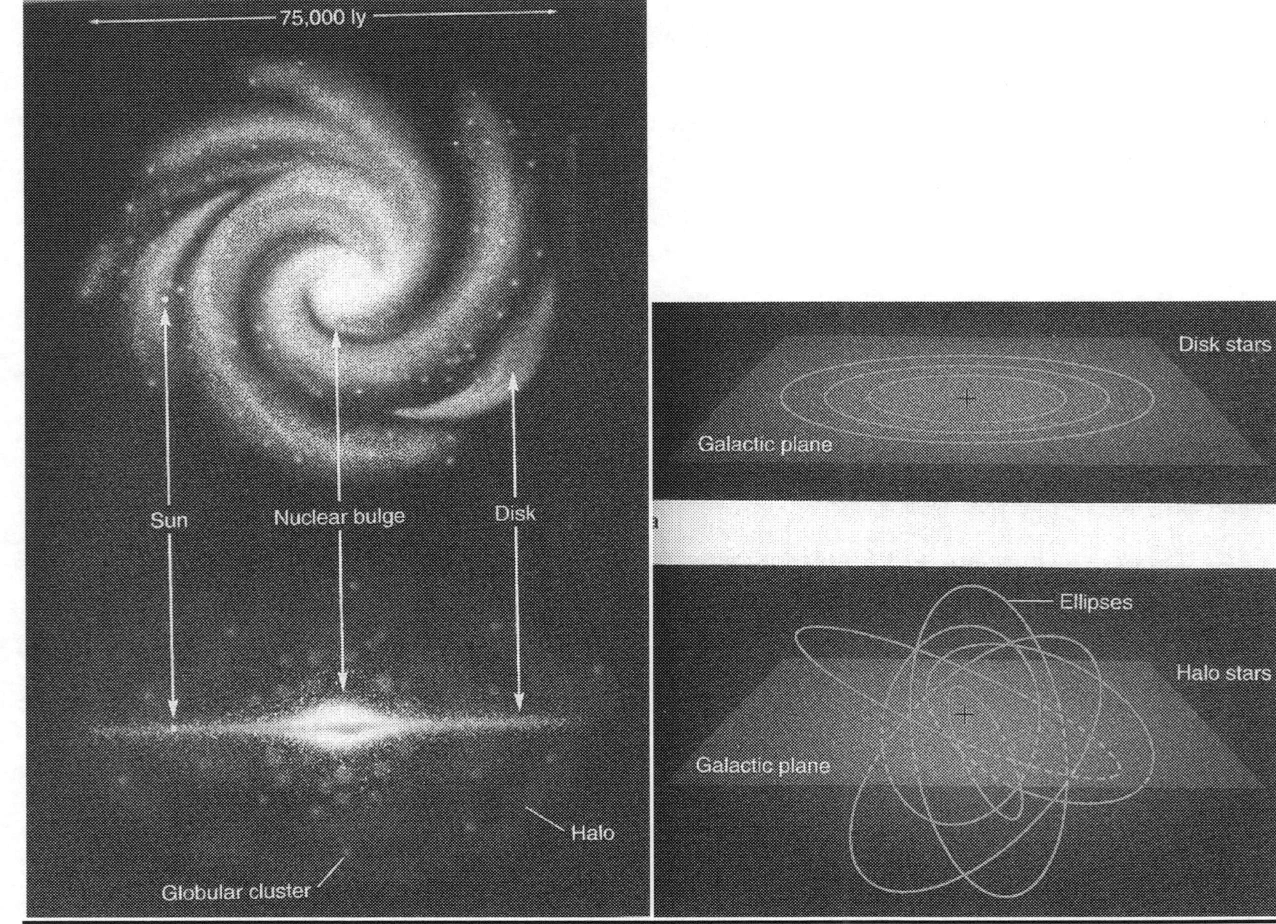

*M.A. Seeds, Horizons: Exploring the Universe, Brooks/Cole –Thompson Learning, 2004, pg. 235

Evidence Supporting Item 7 --- Complicated Star Motions Difficult to Explain

Evidence 7-1 --- Spiral Galaxies – The Milky Way Galaxy

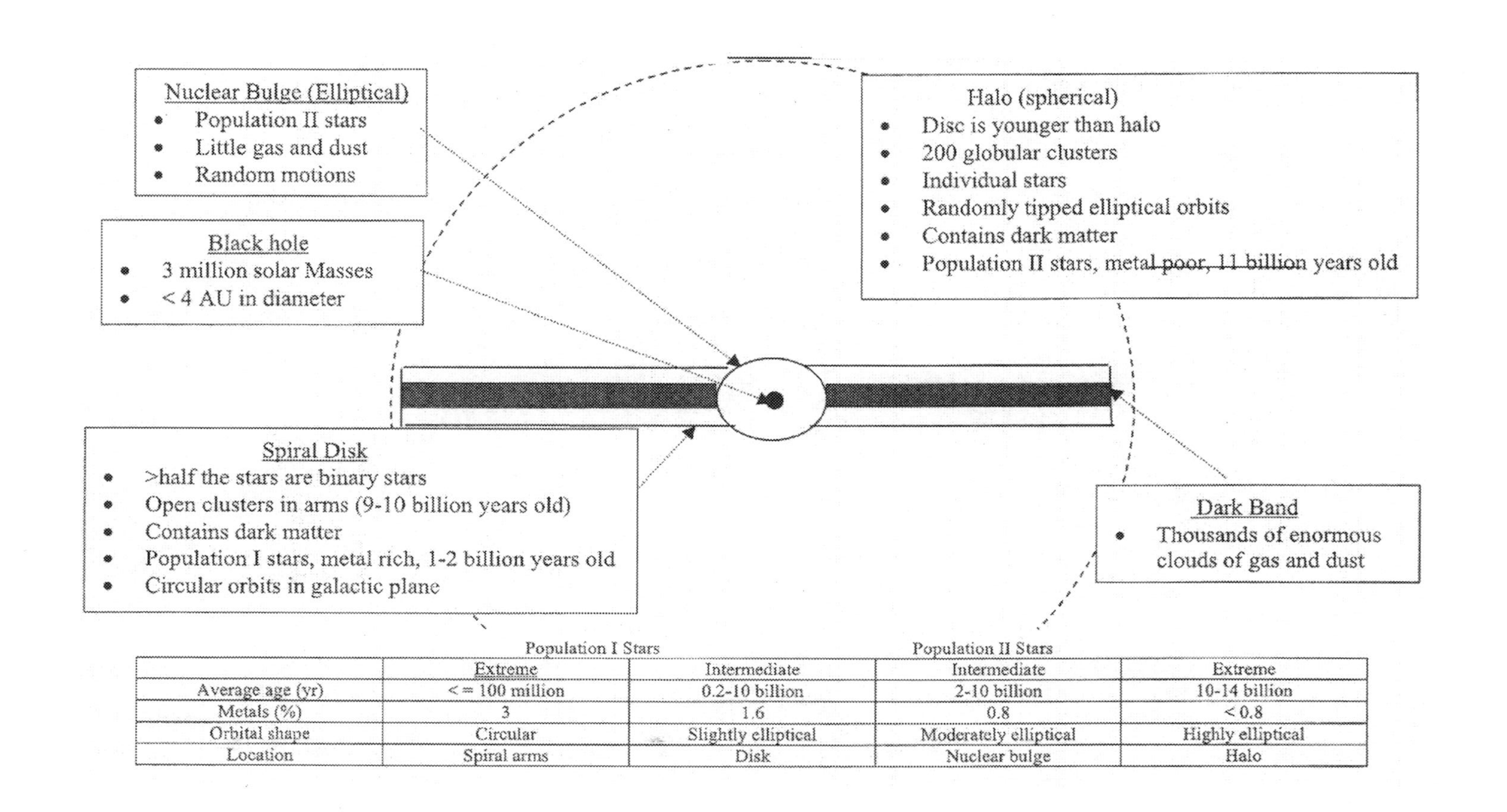

	Population I Stars		Population II Stars	
	Extreme	Intermediate	Intermediate	Extreme
Average age (yr)	<= 100 million	0.2-10 billion	2-10 billion	10-14 billion
Metals (%)	3	1.6	0.8	< 0.8
Orbital shape	Circular	Slightly elliptical	Moderately elliptical	Highly elliptical
Location	Spiral arms	Disk	Nuclear bulge	Halo

(5) REVIEW ASTRONOMICAL OBSERVATIONS THAT SUPPORT TRAVELING-SHOCK-WAVE-PUSHING THEORY

Evidence Supporting Item 7 --- Complicated Star Motions Difficult to Explain

Evidence 7-1 --- Spiral Galaxies – Additional Characteristics

- Most spiral galaxies are believed to have two arms, but some have more
  - There are probably four spiral arms, in the Milky Way Galaxy
- The spiral arms do not rotate like solid bodies
- It takes less time, for stars near the core, to complete one revolution than it does for stars further away from the core
  - Some stars move from one spiral arm to the next
- The gross revolving motion of most spiral galaxies is clockwise, with the tips of the spiral arms trailing
  - Most disk stars revolve with this gross motion, but some disk stars are revolving in the opposite direction
- Some astronomers think they observed spiral galaxies having stars revolving in opposite, counterclockwise direction
- All spiral galaxies are believed to harbor massive black holes
  - The larger the mass of the central black hole, the tighter the spiral arms are wrapped
- Infrared maps of the Milky Way Galaxy reveal that the is a dust ring around a barred core and dust is contained in the spiral arms, bur not between the spiral arms
- The 21-cm emission mapping of gas clouds reveals spiral arms, in addition branches and spurs in the spiral arms

Evidence 7-1 --- Spiral Galaxies – Formation of the Milky Way Galaxy

Traveling-Shock-Wave-Pushing Theory

Three Candidate Scenarios (Ref. 2)

(1) One Shock Wave
(2) Two Shock Waves
(3) Three Shock Waves

Evidence Supporting Item 7 --- Complicated Star Motions Difficult to Explain

Evidence 7-1 --- Spiral Galaxies – Formation of the Milky Way Galaxy

Traveling-Shock-Wave-Pushing Theory -- Basic Scenario Number 3 of 3

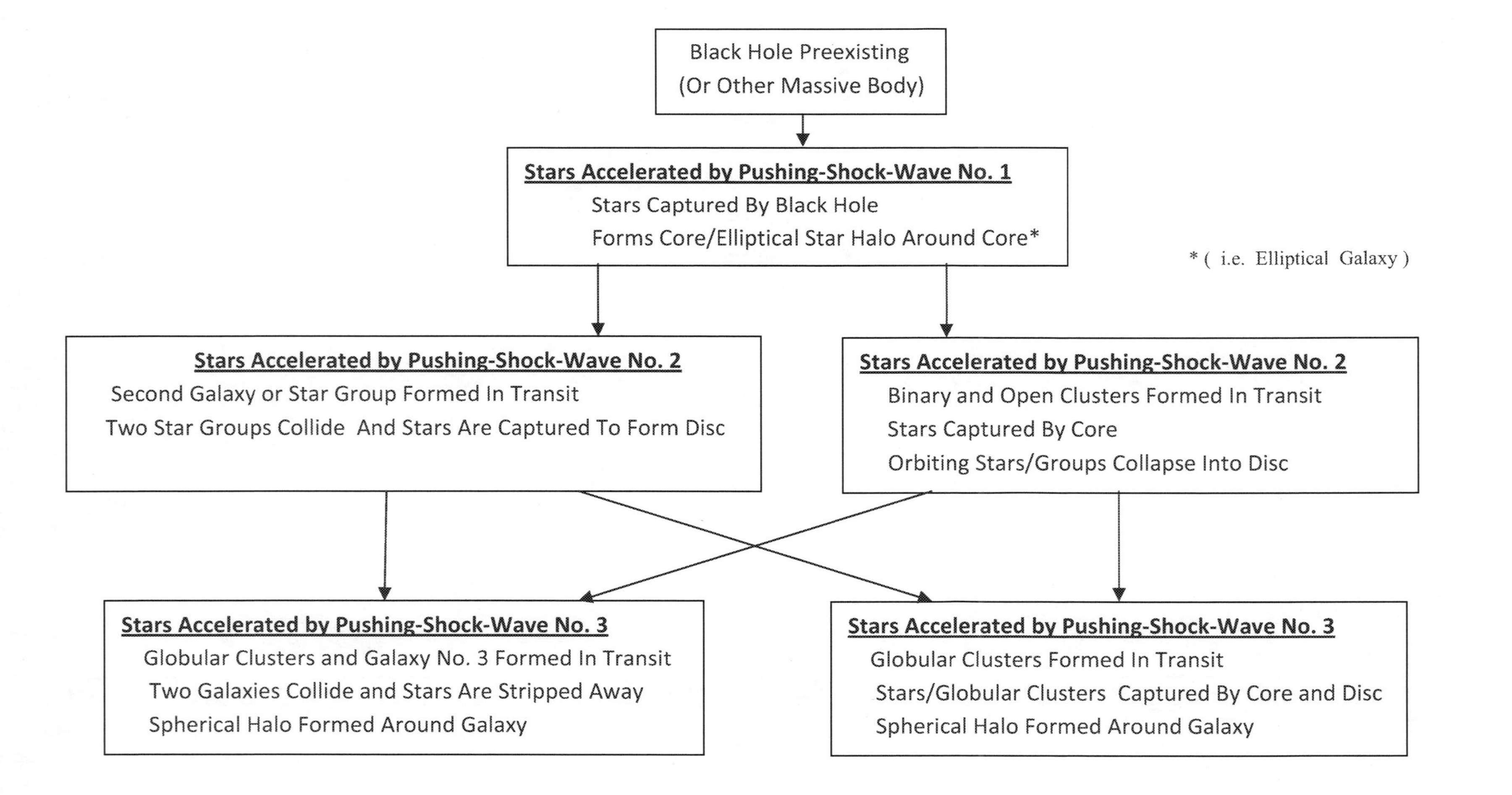

Evidence Supporting Item 7 --- Complicated Star Motions Difficult to Explain

Evidence 7-2 --- Compact Galaxies in Early Universe

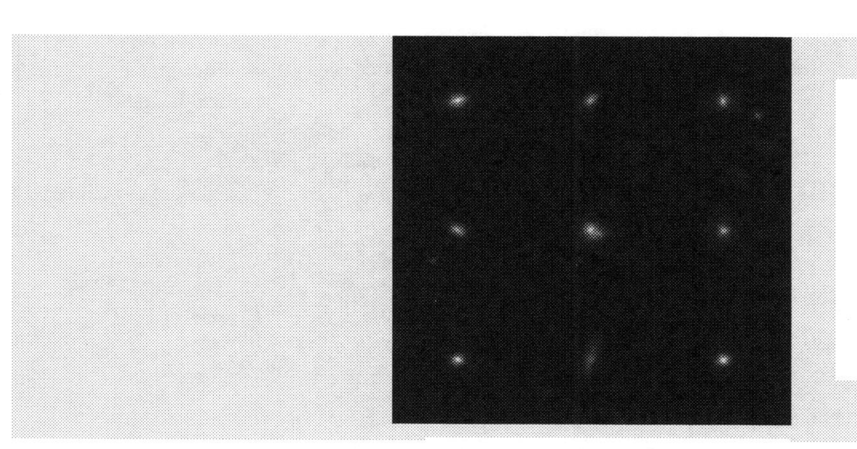

Hubble: STSci-2008-15

- Nine young compact galaxies observed
- Each is 200 billion times the mass of the sun
- Nearly 5,000 light years wide
- **A fraction of the size of today's grownup galaxies**
  - **But, contain the same number of stars**

Traveling-Shock-Wave-Pushing Theory

- A violent explosion occurs at the center of the compact galaxy, creating a residual black hole
- The resulting spherical traveling shock wave pushes the galaxy stars outward, causing the galaxy to enlarge

(6) HOW DARK ENERGY AFFECTS THE HISTORY OF THE UNIVERSE

a. Time_Line From the Big Bang or Rapid Expansion – Candidate No. 1 Scenario

- **Total universe energy concentrated at single point**
- **"Big Bang" or Rapid Expansion" – energy moves rapidly outward**
- **Energy converted to mass**
- **Super massive stars born (short lived)**
  Many other sized stars and dark objects born
- **Super massive stars explode (Hypernovas) creating powerful traveling-shock-waves**
  Traveling-shock-waves push all stars and bodies in their path, clearing spherical volumes (Voids)
- **Accelerated stars are captured and form binary star groups, multiple star groups, open clusters, globular clusters, etc.**
  Galaxy streams are created
- **Star clusters capture other star groups to form disc of spiral galaxies**
  Core-galaxy-star explodes and creates black hole and enlarges galaxy
  Core star of disc galaxy explodes and creates ring galaxy
- **Galaxies capture other galaxies to form revolving galaxy clusters**
  Galaxies collide and alter size and shape of galaxies
  Core/disc of spiral galaxies capture binary stars and globular clusters to create galaxy halo
- **Galaxies move together, from different directions and from different Voids, to form Superclusters, and Walls**
- **All stars are dead. No more explosions to create traveling-shock-wave-pushing of stars**
- **Inertia causes galaxies, stars, etc to continue to expand**
- **Irregular edge of universe – as defined by the location of galaxies – evolves (next page)**

Violent gas ejecting processes create traveling-shock-waves that push stars (shock waves occur throughout space and time)

Star groups and galaxies collide to alter size and shapes of galaxies

b. The Future of the Universe

*Due to Traveling-Shock-Wave-Pushing of Stars

- The edge of the universe – as defined by the distribution of galaxies – will be irregular
- Some of the galaxies – in the furthest reaches of the universe – will be outside observational limits from earth -- due to peculiar velocity
- The Walls will widen, as galaxies from voids on different sides pass through
- The grouping of galaxies, in the labyrinth, will enlarge as galaxies from voids on different sides pass through
- The volume of the Voids will shrink, as walls and galaxy groups expand
- All galaxies and stars will continue to move in a straight line, relative to the expanding universe, at their imparted peculiar velocities – unless they collide with other galaxies

*See next page

b. The Future of the Universe

Due to Traveling-Shock-Wave-Pushing of Stars (See Section 5-4-1)

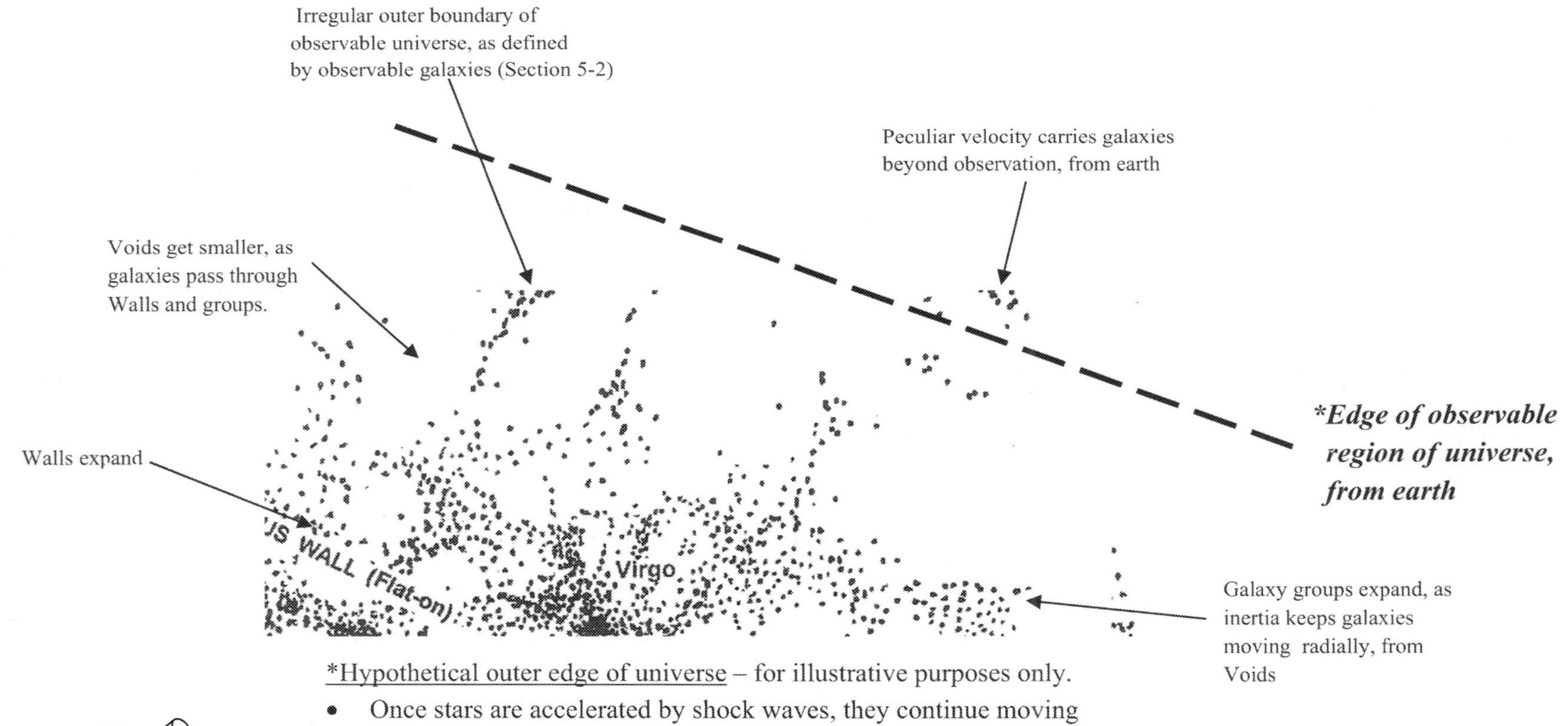

*Hypothetical outer edge of universe – for illustrative purposes only.

- Once stars are accelerated by shock waves, they continue moving in a straight line, unless acted upon by some outside force, causing Walls to expand, Voids to shrink, and galaxy group size to expand

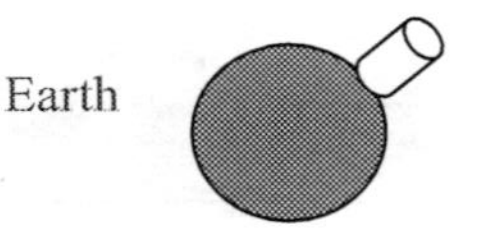

(7) HISTORY OF DARK ENERGY THEORY

STEADY-STATE CHARACTERISTICS
OF A
CROSS-FLOW ELECTRIC ARC

By

AUGUST ALBERT GENKNER JR.

A thesis submitted to the
Faculty of the Graduate School of State
University of New York at Buffalo in partial
fulfillment of the requirements for the degree
of Master of Science

June 1969

March 18, 2004

Dr. August Cenkner, Jr.

Dear Dr. Cenkner,

SUBJECT: PROPOSED THEORY TO EXPLAIN DARK ENERGY

The editors want to thank you for letting them read your article.

Our editorial policy at Sky & Telescope is to report only scientific work which, by appearing in established research journals or professional meetings, has enjoyed some kind of peer review by the research community.

We suggest you send a thorough description of your theory to one of the primary research journals in astronomy or physics. Once your work has been accepted for publication in such a forum, we'd be happy to receive a copy of your refereed article.

Thanks again for taking the time to write.

Sincerely,

Lisa Johnston

Editorial Assistant
Sky Publishing Corp.

49 Bay State Road
Cambridge, MA 02138-1200
USA

Telephone:
617-864-7360

Customer Service:
800-253-0245

Faxes:
617-864-6117 General
617-576-0336 Editorial
617-520-9510 Advertising
617-868-4461 Art & Design

Internet:
info@SkyandTelescope.com
SkyandTelescope.com

Subj: **Astronomical Journal - 204305**
Date: 7/23/04 1:31:36 PM Eastern Daylight Time
From: astroj@astro.washington.edu
To: jandgjr2@aol.com

Dear Dr. Cenkner:

The paper

"Universe Acceleration Predictions using New Dark-Energy Explanations" by August A. Cenkner Jr.

has been reviewed by a knowledgeable referee, whose report is attached.

In view of the referee's report, I believe that the paper is not suitable

for publication in The Astronomical Journal.

Yours Sincerely,
Paul Hodge
Editor

Astronomical Journal
Dept. of Astronomy Phone: 206 685 2150
University of Washington FAX: 206 685 0403
Box 351580 astroj@astro.washington.edu
Seattle WA, 98195-1580 http://www.astro.washington.edu/astroj

---

I have read this paper and strongly recommend that the Editor return it to the author with the indication that its subject matter is not appropriate for the ASTRONOMICAL JOURNAL. It would be better submitted to a journal concerned with fundamental physics and cosmology.

Astronomy®

March 15, 2004

Dr. August Cenkner Jr.

Dear Dr. Cenkner:

Thank you for submitting your article, "Proposed Theory to Explain Dark Energy." Unfortunately, we don't have a place for it in our packed editorial schedule.

As per our standard procedure we've enclosed the article you sent. Thanks again.

Sincerely,

Dave Eicher
Editor

21027 Crossroads Circle, P.O. Box 1612, Waukesha, WI 53187-1612
phone 262.796.8776
website: www.astronomy.com

NEWSLETTER OF THE BUFFALO ASTRONOMICAL ASSOCIATION INC.

The Spectrum

Table of Contents

A Theory to Explain Dark Energy – Part I 1
BAA Officials 2
BAA Web Site 2
Meeting Location/Time 2
Spectrum Deadline 2
Presidents Message 2
Upcoming Meetings 3
Convince Us, If You Can 3
College of Fellows Report 3
Funny, Funny, Funny 3
BAA Annals 3
Catastrophe, Catastrophe, Catastrophe 4
Spy and Tell 5
Astronomy Terms 5
Observatory Notes 7
BAA Policy 7
BAA Considers Joining Search for New Planets 8
Cartoons 8
A Dark-Energy Theory – Part II 9
Editors Corner 10

A THEORY TO EXPLAIN DARK ENERGY – PART I

The Accelerating Universe

Dr. August Cenkner Jr. Copyright March 7, 2004

Astronomers have observed that stars and galaxies, at the outer reaches of the universe, have actually accelerated and are therefore moving at higher, rather than the anticipated lower, velocities. Classical thought is that gravitational attraction, between celestial objects, would actually slow them down instead of speeding them up. They have attributed this unexpected behavior to some mysterious "dark energy", that apparently permeates the universe. The nature of this dark energy is presently unknown.

A dark energy theory is presented herein, that could explain the observed behavior of these astronomical bodies.

The basic theory states that multiple hypervelocity traveling gas-pressure/radiation-pressure/stellar-wind shock-waves are producing numerous force impulses when they impinge on an astronomical body. If the body is in front of the shock-emitter, these impulses would cause it to speed up. If it is behind the shock-emitter, it would slow down. The shock waves would be moving at a velocity that is higher than the emitter velocity, which would enable the shock to overtake bodies moving in front of the emitter. Bodies traveling behind the emitter would slow down because of retarding shock impulses.

This behavior can be predicted by applying Newton's Law of Motion. The unbalanced pressure on a star, that occurs upon shock impact, will produce a force impulse [force x time] that will increase the linear momentum [mass x velocity] of the astronomical body and hence increase its velocity and acceleration.

These traveling shock waves will also affect the velocity and path direction of nebula, (and possibly stimulate star formation) as these waves pass through the nebula. Nebula that are in front of the shock emitter would be accelerated while nebula that is behind it would be decelerated. Since there could be more than one shock-emitter in a galaxy or star cluster, these multiple shock waves would interact. The shocks could be reflected, attenuated or enhanced and could produce multiple or enhanced shock impulses.

The so-called dark energy is actually the energy contained in these high energy hypervelocity traveling shock waves. The thermal kinetic and internal energy of the star is converted into the kinetic and internal energy of the traveling shock wave. When the shock impinges on a star, or other body, it performs work on the body. This work increases the kinetic energy of the body.

Possible sources of these shock waves include exploding stars, pulsating stars, imploding stars, colliding stars, black holes as matter approaches the event horizon, stellar wind, etc. This shock wave energy would permeate the entire universe over time, since these shock-emitters would appear, eventually, wherever stars exist.

It has been reported that the core of the universe doesn't presently exhibit the presence of dark energy. Rather, it occurs in an expanding spherical shell around the core. This suggests that the stars are still aging in the core and they have to reach a certain level of maturity before they can become shock-emitters. Once they reach maturity in the dark energy zone, they are transformed into shock-emitters. This seems to suggest that star explosions and collapsing of mature stars may be the major emitters, with supernova and cataclysmic variables being the prime candidate as the most powerful of the relevant shock-emitters.

The theory suggests that, eventually, all shock-emitters will have been consumed. Gravity will again become the dominant force and drag may more significantly influence cosmic motion. The expansion of the universe will slow down, perhaps to the point that the universe eventually collapses upon itself.

Of course, some of the remnants of the shock-emitters may eventually coalesce into a new generation of stars. Some of these stars may then eventually transform into a new generation of shock-emitters. This would enlarge the dark energy zone and prolong expansion. *(continued on page 9)*

Subj: **DARK2004 Conference Schedule**
Date: 9/20/04 3:41:43 PM Eastern Daylight Time
From: b-guster@physics.tamu.edu
To: b-guster@physics.tamu.edu

Dear Sir,

This email is to let you know that we are unable to accommodate your request to make a presentation at the upcoming DARK2004 Dark Matter in Astro and Particle Physics Conference, October 4-9, 2004. The conference is shaping up to be quite packed with invited speakers and so we are unable to add any others, however, we do appreciate your interest and hope that you can attend. I have attached a preliminary schedule for you.

Again, thank you for your interest.

Sincerely,

Beverly Guster

Beverly Guster, Program Coordinator
Mitchell Institute for Fundamental Physics
4242 TAMU
College Station, TX 77843-4242

Phone (979)845-7778
FAX (979)845-8674
b-guster@physics.tamu.edu

UNIVERSE ACCELERATION PREDICTIONS USING NEW DARK-ENERGY EXPLANATION

August A. Cenkner Jr., Ph.D.

Buffalo State College, Elmwood Avenue
New Science Building
First Floor Auditorium, Room 213

Friday, September 10, 7:30 P.M.

In 1998, by studying the emission spectra from galaxies in the outer reaches of the universe, two independent teams of astronomers concluded that these galaxies have actually accelerated; they are moving at higher, rather than the anticipated lower, velocities (Ref 1-4). Classical thought is that gravitational attraction, between celestial objects, should actually slow them down instead of speeding them up. They have attributed this unexpected behavior to some mysterious repulsive force that apparently permeates the universe; it has been labeled dark-energy. The nature of this dark-energy is presently unknown. In addition, there are other currently unexplained phenomena like revolving star/dark body groups, galactic superclusters, voids, walls and nebula vortex. These phenomena will also be discussed and related to dark-energy. Finally, dark-energy will be exposed as a gas dynamic phenomenon and it will be shown how star acceleration occurs. This includes the results of laboratory simulations of dark-energy, that were performed at the University at Buffalo.

REFERENCES

(1) Villard, R. & Lloyd,R., **"Astrophysics Challenged by Dark Energy Finding"**, Space.com, April 2001

(2) Chaikin, A., **"Dark Energy: Astronomers Still Clueless About Mystery Force Pushing Galaxies Away"**, Space & Science, 2002.

(2) Weinberg, S., **"Importance of Discovering the Nature of Dark Energy"**, Dept of Physics, University of Texas at Austin

(4) Krauss, L. M., **"Dark Energy and the Hubble Age"**, Astrophysics Journal, April 1, 2004

September 4, 2004

Professor Dale Taulbee
Department Chairman
Mechanical And Aerospace Engineering
315 Jarvis Hall, North Campus
University at Buffalo
The State University of New York
Buffalo, NY 14260

Dear Professor Taulbee:

I was inspired, by basic experimental research that I did in the School of Engineering at UB, more than 30 years ago, to formulate a theory to explain some here-to-fore unexplained astronomical phenomena. Also, all the gas dynamics education that I received at UB, including many courses that you taught me, gave me the insight to recognize the cause of these unexplained phenomena.

For years now, astronomers have been unable to come up with an explanation f or these phenomena. I suspect that it's because many of them do not have extensive backgrounds in gas dynamics.

My gas dynamic theory is consistent with astronomical observational data and basic research that I performed with Dr. David Benenson, while working on my M.S. and Ph.D. degrees.

If I'm right - and of course I think I am – this is a major breakthrough in theoretical astrophysics.

I just thought that the School of Engineering might be interested in learning about this.

Yours truly

Dr. August Cenkner Jr.
Class of 66, 69, 73, 96

November 4, 2005

Subject: **Star Acceleration and Dark Energy, Correlated with Small Scale Laboratory Simulations**

Memo To: Supernova Cosmology Project

Greg Aldering
Susana Deustua
Gerson Goldhaber
Ariel Goobar
Don Groom
Isobel Hook
Dan Kasen
Alex Kim
Rob Knop
Peter Nugent
Reynald Pain
Saul Perlmutter
Carl Pennypacker
Michael Wood-Vasey

High-Z Suoernova Search Team

Brian P. Schmidt
Nicholas B. Suntzeff
Robert Schommer
Chris Smith
Mark M. Phillips
Alejandro Clocchiatti
Bruno Leibundgut
Jason Spyromilio
John L. Tonry
Alex V. Filippenko
Weidong D. Li
Adam G. Riess
Craig J. Hogan
Chris Stubbs
Peter M. Garnavich
Stephen Holland
Robert Kirshner
Saurabh Jha
Thomas Matheson
Peter Challis

Universitv of Texas at Austin
Professor Steven Weinberg

Universitv of Rochester
Professor Judith Pipher
(May also explain the "wild star" (i.e. the older age star in a newer age star zone) that Spitzer discovered)

SNAP Team Members
(Supernova/ Acceleration Probe -New satellite for dark energy research)

Universitv at Buffalo
Professor Will Kinney
Professor B.A. Weinstein

Buffalo State College
Dr. Jack Mack

BAA Board Of Directors

Peter Proulx
Ted Bistany
Joe Orzechowski
Bev Orzechowski
Paul Tabor
Janice Gardner
Tom Bakowski
Richard Fusani
Bill Aquino
Rowland Rupp

From: August Cenkner Jr.
B.A., B.S., M.S., Ph.D., University at Buffalo, Buffalo, New York

Summary:

The **enclosed book**, " A Dark Energy Theory Correlated with Laboratory Simulation and Astronomical Observations (first edition)", introduces a new classical theory for dark energy, and shows how stars can accelerate. The theory is guided by small scale laboratory simulation of dark energy, as it interacts with a main sequence star.

Subj: **JCAP/012A/0804**
Date: 8/20/04 4:54:26 AM Eastern Daylight Time
From: jcap-eo@jcap.sissa.it
To: jandgjr2@aol.com

Dear Dr. Cenkner,

we regret to inform you that your paper "Universe Acceleration Predictions Using New Dark Energy Explanation" is not suitable for JCAP and has just been withdrawn from our journal.

Best regards,
JCAP Executive Office

JCAP Executive Office - http://jcap.sissa.it
SISSA, Via Beirut 2-4, 34014 Trieste (Italy)
tel +39-040-3787571, fax +39-040-3787528

THE ASTROPHYSICAL JOURNAL

Logged in: August Cenkner (jandgjr2@aol.com)

AUTHORS: Check MS status

MS 61394

Title	Dark Energy Theory, with Laboratory Simulation, to Explain Universe Acceleration, Superclusters, Voids, Walls, Revolving Star/Dark Body Groups, Drag, and Wild Stars
Author(s)	August Cenkner, Jr.
Corresp. author	August Cenkner (jandgjr2@aol.com) [Change]
Proofs author	August Cenkner (jandgjr2@aol.com) [Change]
Editor	()
Subjects	These subjects have **not** been verified by the journal office: cosmology:theory-cosmology:miscellaneous-large scale structure of universe-shock waves

Versions

Version	Received	
1	29 September 2004	Access MS files | Upload additional files

Decision history

Version	Decision	Date
1	no decision yet	

Journal and Supplement: apj@as.arizona.edu Tech support: apj-help@mss.uchicago.edu
Letters: apjletters@letters.as.utexas.edu

W3C WAI-AA WCAG 1.0 508 BOBBY APPROVED

Web site: dark-energy-theory.com

A DARK ENERGY THEORY CORRELATED WITH LABORATORY SIMULATIONS AND ASTRONOMICAL OBSERVATIONS

Book Written by :

August A. Cenkner Jr .

B.A., B.S., M.S., Ph.D.

University at Buffalo

Buffalo, New York, USA

In 1998, by studying the emission spectra from galaxies in the outer reaches of the universe, two independent teams of astronomers concluded that these distant galaxies have actually accelerated --in contradiction with classical expectations. They have attributed this unexpected behavior to some mysterious unknown repulsive force, that they have labeled dark energy.

Introduced herein is a classical theory for this elusive dark energy -- the mysterious repulsive force that is causing stars, in the outer reaches of the universe, to accelerate. Once dark energy is revealed, an application of Newton's Law of Motion is used to first demonstrate how this acceleration occurs. It is then used to show that dark energy may also be responsible for other unexplained astronomical phenomena like galactic Clusters, Voids, Walls, and revolving star/dark-body groups. Beyond this, it is shown that dark energy can, as well, be responsible for drag on stars and dark bodies, nebula vortex, and "wild stars and/or wild dark-bodies".

Small-scale simulation of dark energy, in a plasma laboratory, is used to guide the evolution of the dark energy theory and to demonstrate what happens when dark energy begins influencing main sequence stars.

Theoretical predictions are revealed to be consistent with reported astronomical observations.

A DARK ENERGY THEORY
CORRELATED WITH
LABORATORY SIMULATIONS
AND
ASTRONOMICAL OBSERVATIONS

(first edition)

Paperback (6x9)

order:

www.AuthorHouse.com

March 13, 2005

Letters To the Editor
Sky & Telescope
49 Bay State Rd.
Cambridge, Ma 02138
Letters@SkyandTelescope.com

Dear Sir/Madam:

I would like to submit the following comments for consideration for publication in Letters.

Mr. Sean Carroll recently noted in his article (March 2005, page 32), "Dark Energy & The Preposterous Universe", that astronomers have still been unable to identify what dark energy is and how it causes some distant galaxies to accelerate. In the recently published book , "A Dark Energy Theory correlated with Laboratory Simulations and Astronomical Observations", a plausible classical explanation is given for dark energy. The explanation is used to predict main sequence star accelerations, using Newton's Law of Motion., and to show how dark energy could have also created Clusters, Voids, and Walls (as Mr. Carroll speculated it may have). It is also shown that dark energy may account for other unexplained phenomena like revolving binary stars, revolving star clusters, counter rotating galaxies, and galactic collisions or close encounters that are the result on one galaxy overtaking and passing another galaxy. Finally, it is shown that under certain conditions dark energy will create wild-stars and wild-dark-bodies; i.e. free floating bodies that have radial velocities that are opposite to the direction of expansion of the universe. The explanation is guided by small-scale laboratory simulations of dark energy, in a plasma tunnel, as it starts interacting with a main sequence star. The explanation is shown to be consistent with astronomical observations.

August Cenkner Jr.
B.A., B.S., M.S., Ph.D.
DarkEnergyTheory@aol.com

Thank you for your consideration

Dr. August Cenkner Jr.

Subj:	**Autoreply from SKY & TELESCOPE**
Date:	4/2/2005 3:54:07 PM Eastern Standard Time
From:	letters@SkyandTelescope.com
To:	jandgjr2@aol.com

Sent from the Internet (Details)

Thank you for writing to SKY & TELESCOPE. We have received your letter and will consider it for publication in an upcoming issue of the magazine. Because of the large volume of mail we receive, not all letters will be published, nor can they all be answered personally.

If you didn't mean to submit your letter for publication but were intending to correspond with our Editorial Department for some other reason, please resend your e-mail to our general editorial address: editors@SkyandTelescope.com.

Thank you for your interest in SKY & TELESCOPE. Clear skies!

-- The Editors

March 10, 2006

UB Today
University at Buffalo
330 Crofts Bldg.
Buffalo, NY 14260
whitcher@buffalo.edu

Dear Sir/Madam:

I would like to submit my recently published book for consideration for inclusion in "read&listenON!". Attached is a copy of the book cover.

Authors Degrees from UB:	BS -- Aerospace Engineering	1966
	MS -- Engineering Science	1969
	PhD -- Engineering Science	1973
	BA -- Computer Science	1996

Publisher: Authorhouse

Publication: August 23, 2005

Background: Involved in Research and Development, in relevant areas, for 34 years. Has authored or co-authored 21 publications. Taught for 11 years in the UB School of Engineering, for Millard Fillmore College. Currently retired.

Book Title: A Dark Energy Theory Correlated with Laboratory Simulation and Astronomical Observations

Laboratory Simulation: Performed in the Plasma Physics Laboratory, at the UB School of Engineering.

Description of the Book

In the astronomical area, there are still many unexplained phenomena. For example, by studying the emission spectra from galaxies in the outer reaches of the universe, in 1998, two independent teams of astronomers have concluded that these galaxies have actually accelerated. This contradicts expectations that they should have decelerated, due to gravitational attractions with other bodies. This unexpected behavior has been attributed to some mysterious repulsive force that has been named dark energy.

Introduced herein, for the first time, is a classical face for this elusive dark energy. The prediction of star acceleration is evolved from research conducted at the University of Buffalo in 1969, for an unrelated project. It is shown that this research can be reinterpreted as a small-scale simulation of dark energy, as it interacts with a main sequence star. This leads to an application of Newton's Law and a prediction of star acceleration.

It is then shown how dark energy can also be responsible for other unexplained phenomena like Voids, Walls, Clusters and groups of revolving star/dark-bodies.

Yours truly
August Cenkner Jr.

A S T R O N O M I C A L S O C I E T Y O F T H E P A C I F I C
390 Ashton Avenue ❖ San Francisco, CA 94112 USA ❖ Tel (415) 337-1100 ❖ Fax (415) 337-5205 ❖ www.astrosociety.org

13 March 2006

August Cenkner

Dear Mr. Cenkner:

Thank you for your request to place advertising in the next issue of Mercury. As you may be aware, *Mercury* is the membership magazine of the Astronomical Society of the Pacific, and very little space is reserved for advertising. Indeed, we often decline ads for this reason. In your case, I am sorry to say we must decline your ad. We do wish you well with your new book, however. Your CD and check are returned herewith.

Sincerely,

James C. White II
Editor

Ad in Sky & Telescope	
Pg 94	July 2006

A Dark Energy Theory
Correlated With
Laboratory Simulations and Astronomical Observations

August A Cenkner Jr
BA, BS, MS, PhD

A classical face is placed on the very elusive dark energy - the mysterious repulsive force that is causing stars in the outer reaches of the universe to accelerate, instead of decelerate due to gravity. Laboratory simulation of dark energy in a plasma tunnel, as it interacts with a main sequence star, is used to guide the creation of the theory. This leads to the application of Newton's Law of Motion, to predict star acceleration. It is shown that dark energy can also be responsible for other unexplained phenomena like Clusters, Voids, Walls and revolving star/dark-body groups.

Barnes & Noble Amazon.com

ISBN 1-4208-3447-9

THE ASTROPHYSICAL JOURNAL

ROBERT C. KENNICUTT, JR., *Editor-in-Chief*
Steward Observatory
University of Arizona
933 North Cherry Avenue
Tucson, Arizona 85721-0065
Telephone: *520-621-5145*
Facsimile: *520-621-5153*
E-Mail: apj@as.arizona.edu

January 31, 2005

Dr. August Cenkner

Dear Dr. Cenkner:

I am writing to you with regard to your manuscript entitled "Dark Energy Theory, with Laboratory Simulation, to Explain Universe Acceleration, Superclusters, Voids, Walls, Revolving Star/Dark Body Groups, Drag, and Wild Stars", which you recently submitted to The Astrophysical Journal. I have read your manuscript, and considered its appropriateness for publication in our journal. Unfortunately, the paper does not make a sufficiently convincing case to merit publication in our journal.

Consequently, I regret to inform you that we will be unable to publish your article. Nevertheless, I do offer you my best wishes in your future research.

Sincerely,

Robert C. Kennicutt, Jr.
Editor-in-Chief

RCK:teb

Published by The University of Chicago Press, 1427 E. 60th Street, Chicago, Illinois 60637
for THE AMERICAN ASTRONOMICAL SOCIETY

UB today

UNIVERSITY AT BUFFALO

Read & Listen On

Recent Books by UB Alumni

Hurricane Katrina: Response and Responsibilities

Edited by John Brown Childs, PhD '75

Baseball and the Music of Charles Ives: A Proving Ground

By Timothy A. Johnson, PhD '91

A Dark Energy Theory Correlated with Laboratory Simulation and Astronomical Observations

By August Cenkner, BA '96, PhD '73, MS '69 & BS '66

By studying the emission spectra from galaxies in the outer reaches of the universe, two independent teams of astronomers in 1998 concluded that the universe is expanding at an ever-faster pace. This contradicts expectations that the universe's expansion was decelerating, due to gravitational attractions with other bodies. This unexpected behavior has been attributed to some mysterious repulsive force that has been named dark energy. Also shown is how dark energy may be responsible for other unexplained phenomena like voids, walks, clusters and groups of revolving star/dark bodies. (*AuthorHouse, 2005*)

Search results for "August Cenkner"

search

August Cenkner 

search in

- web
- pictures
- video
- audio
- news
- local
- shopping

recent searches

August Cenkner

web results

UB Today: Read & amp; Listen On
By **August Cenkner**, BA '96, PhD '73, MS '69 & BS '66. By studying the emission spectra from galaxies in the outer reaches of the universe, two independant ...
http://www.buffalo.edu/UBT/UBT-archives/32_ubtss06/read_listen/

Cenkner Jr, **August** A.: A Dark Energy Theory Correlated with ...
Cenkner Jr, **August** A.: A Dark Energy Theory Correlated with Laboratory Simulations and Astronomical Observations,Conservative,Book Club,Reagan,Homeschool ...
http://www.forbesbookclub.com/bookpage.asp?prod_cd=IUF17

Buffalo Astronomical Association » News
BAA member Dr **August Cenkner** Jr. recently published a research book entitled : " A Dark Energy Theory Correlated With Laboratory Simulations And ...
http://www.buffaloastronomy.com/index.php?id=24

Flickering and Periodic Activity in the 2004 Outburst of BZ UMa
5 **August**. Joe Orzechowski. To Be Announced. 8:30 pm. 9:34 pm. Waxing crescent (2% illuminated). 20 **August**. Gus **Cenkner**. Visual Tour Of TheUniverse ...
http://www.buffaloastronomy.com/pdf/SpectrumV7I4_JulyAugust2005.pdf

Barnes & Noble.com Books - Authorhouse
August Cenkner / Paperback / Authorhouse / May 2005 Our Price: $13.45 Usually ships within 2-3 days. 7. Book Cover · Tale of the Lobster: A Practical Guide ...
http://btobbrowse.barnesandnoble.com/browse/ ... , 141443637&Ne=164605&z=y&btob=Y

Caroline Herschel COLLEGE OF FELLOWS MEETING
August Cenkner Jr., Editor. 6982 Creekview Drive, Pendleton, New York. Lockport PO 14094. BAA Member Reviews Astronomy Course. Gus **Cenkner** ...
http://www.gotastronomy.com/Spectrum/Spectrum-Jan-Feb-vol-8-1.pdf

BAA News Book Review
August Cenkner Jr., Editor. 6982 Creekview Drive. Pendleton, NY. Lockport PO 14094. Page 6. The Spectrum - May/June 2006. Volume 8 Issue 3 ...
http://www.gotastronomy.com/Spectrum/Spectrum-May-June-vol-8-3.pdf

Table Of Contents (condensed)
August A. **Cenkner** Jr. BA, BS, MS, Ph.D. University at Buffalo. Buffalo, New York. USA. In 1998, by studying the emission spectra from galaxies in the outer ...
http://www.dark-energy-theory.com/

Finding a Way to Test for Dark Energy
August 30, 2005 - BERKELEY, CA - What is the mysterious dark energy that's causing the expansion of the universe to accelerate? ... by **August** A. **Cenkner** Jr ...
http://www.brightsurf.com/news/headlines/view.article.php?ArticleID=20765

9/9/2006

Amazon.ca Books: Dark energy (Astronomy)
A Dark Energy Theory Correlated with Laboratory Simulations and Astronomical Observations by **August** A. **Cenkner** Jr (Paperback - May 2006) ...
http://www.amazon.ca/s?ie=UTF8&index=books-c ... %20energy%20 (Astronomy)&page=1

1 2 3 4 5 6 7 8 9 10

keep searching web for:

August Cenkner

search

"Origins of Dark Energy"

Conference

May 14 –17th, 2007
McMaster University, Hamilton, Ontario, Canada

This international conference on DARK ENERGY is jointly organized by the Origins Institute (0I) at McMaster University and the Perimeter Institute (PI), co-sponsored by the Canadian Institute of Theoretical Astrophysics (CITA).
The conference portion of this meeting will be hosted by the Origins Institute for 4 days --Monday May 14 through Thursday May 17 at the campus of McMaster University, in Hamilton Ontario. The conference will be immediately followed by a workshop to be held from Friday May 18 through Sunday May 20 at Perimeter Institute for Theoretical Physics (PI) in nearby Waterloo, Ontario.

The conference is meant to bring together observers and theorists in astronomy, cosmology and particle physics ... observational evidence and theoretical ideas for Dark Energy, and to highlight the most promising future directions. ..... meeting includes broad review talks with which to start the discussion on each of the main areas of enquiry .

The workshop will focus novel theoretical ideas on the nature of the dark sector and their prospects for observation .... format is intended to be very informal, with few talks and ample time for discussions and interactions.
The main topics to be covered during the conference are:

I OBSERVATIONAL EVIDENCE FOR DARK ENERGY

1.CMB
2. SN/Weak Lensing
3. large Scale Structure

II THEORETICAL APPROACHES

4. Cosmology-Driven Approaches
5. IR Modifications to Gravity
6. Vacuum Issues and the Cosmological Constant Problem
7. More Radical Approaches

8. Future Directions

http :/ /origins.mcmaster .ca/ darkenergy /ab out. html 2/18/07

Subj: Re: Presentation
Date: 4127107 1 :01 :53 PM Eastern Daylight Time
From: ori ins univmail.cis.mcmaster.ca
To: DarkEnergyTheoly@aol.com

Hi Dr. Cenkner

You may bring a poster and there will be poster boards provided at the conference for you and others. The webmaster is not putting the poster information on the web. We are only putting the title and abstracts of Invited Talks and contributed talks on the web.

Thanks, Rosemary

On Wed, 25 Apr 2007 15:52:21 EDT DarkEnergyTheory@aol.com wrote:
➢ I volunteered to give a presentation at the Dark Energy Conference,
➢ but I was notified that there wasn't any available time. Instead, you
➢ said I should submit a poster presentation. So I submitted an abstract
➢ for a poster presentation.
➢ I never got a response and my name doesn't appear on your
➢ website list of presenters.
> Could you let me know what's going on, so I can get prepared.
> Thank you for your assistance.
> Dr. August Cenkner
➢ ************************************ See what's free at
➢ *http:llwww.aol.com.*
➢ ******* **** ** * ** ** **** ** ** *** ** ** ** ** ** ** *** ** ** **** **

Rosemary McNeice, Origins Secretary
McMaster University
ABB-241, 1280 Main Street West,
Hamilton, ON L8S 4M1
905-525-9140 X23531

Page 1 of 1

Rosemary

I don't think it's fair or proper for an institution of higher learning to censor what is presented at a conference.

I have spent years conducting relevant original theoretical research that the scientific community should have a right to be made aware of and to publicly debate.

It's certainly highly improper for a webmaster to have the power to decide what research is relevant and what isn't.

At every conference that I've attended, the participants have always been made aware of everything that was to be presented at the conference. That's why we attend the conferences.

Dr. August Cenkner

Saturday, April 28, 2007 America Online: DarkEnergyTheory

Subj: (no subject)
Date: 511107 5:35:49 PM Eastern Daylight Time From: cburgess@Qerimeterinstitute.ca To: DarkEnerayTheo!y@aol.com
CC: origins@univmail.cis.mcmaster.ca, kgillesQie@Qerimeterinstitute.ca, cburgess@Qerimeterinstitute.ca

Dear Dr. Cenkner,

I recently was forwarded your message regarding posters, and please do prepare a poster for our conference which you can display at the poster session which we will keep open where people have coffee near the
lectures. This encourages people to come view them in the breaks between sessions.

Unfortunately we have no plans for posting the poster titles on the web page. Posting poster titles is not a common practice in my area, and so we did not set up the web page to allow us to do so. (Of course this is not specific to you, but is true for all of the poster submissions.)

Best regards, cliff burgess

Thursday, May 03,2007 America Online: DarkEnergyTheory

Subj: Re: Conference Dispute

Date: 5/4/07 9:43:40 AM Eastern Daylight Time
From: DarkEnergyTheory
To: cburgess@pcerimeterinstitute.ca
CC: origins univmail.cis.rncmaster.ca, kgillespie@perimeterinstitute.ca

Dear Cliff Burgess

A considerable amount of time, money, and effort was expended on my research project. I therefore find your cavalier attitude, about my work, to be outrageous and extremely insulting.

Your decision to censor me from the dark energy conference web site means that everybody who visits the web site, and many participants, will not even be aware of my research. You have essentially decided that my work is insignificant and of no consequence.

A university conference is supposed to be set up for the free exchange of ideas. You have made a mockery of this worthy goal by deciding which work is relevant and which isn't; this decision should be made by the scientific community.

It's pretty obvious to me that you could have posted my abstract, along with all the one's you already posted, if you really wanted to.

You certainly have succeeded in establishing a reputation for yourself and for McMaster University.

It seems logical to conclude that we can expect the same type of treatment at the upcoming workshop.

Dr August Cenkner

See what's free at AOL.com.

Friday, May 04, 2007 America Online: DarkEnergyTheory

Subj: Re: Conference Dispute
Date: 514107 9:49:07 AM Eastern Daylight Time From:
cburgess@Qerimeterinstitute.ca To: DarkEnergyTheory@aol.com

CC: origins@univmail.cis.mcmaster.ca, kgillesQie@~rimeterinstitute.ca, cburgess@Qerirneterinstitute.ca, jkhoury@Qeri rneterinstitute.ca , Qudritz@thyme .P hvsics.McMaster .CA

Dear dr cenker,

I'm sorry you do not wish to have the same opportunity to present your results as do all the other participants. If you wish to withdraw from the conference I would understand completely.

Best regards, cliff

Sent from my BlackBerry Wireless Handheld

"Origins of Dark Energy"

Conference

Author's Comments

- While at the conference, I gave out more than fifty (50) free copies of the first edition of this book. Included was a copy of the cover of the second edition, since it wasn't published at that time.
- I mounted a presentation on a bulletin board in the hallway. I used many of the slides included in the first edition and some from this second edition.
- After the first presentation by Dr. Ed Copeland, on "Dynamics of Dark Energy", the floor was opened to the audience. At that time, I introduced my theory -- that dark energy is actually the energy contained in a traveling shock wave. I explained the laboratory simulation that I had performed, along with my interpretation. I also noted that I thought dark energy is responsible for the formation of spiral galaxies and other star groups, in addition to the creation of voids, walls and clusters.

  I concluded by noting that I thought any parameters, in theoretical models of dark energy, should be space and time dependent -- because the shock waves are localized in space and time. Most, if not all of the conference participants, appeared to be present, in addition to many students and faculty.
- There were three other (short) poster presentations. Some of the speakers that were listed in the conference program did not make a presentation.
- The stated objectives, of the conference, were advertised to include:

  (7) More Radical Approaches

  (8) Future Directions
- Since I wasn't allowed to give an oral presentation, participants lost any chance at asking questions and at meaningful brainstorming.

Attendee's Name Tag

May 31, 2007

TO: Dakila D. Divina	Editor
Managing Editor	The Buffalo News
Parade Publications	One News Plaza
711 Third Street	P.O. Box 100
New York, N.Y. 10017	Buffalo, N.Y. 14240

FROM: Dr. August Cenkner Jr.

SUBJECT: PROPOSED FOLLOW-ON DARK ENERGY ARTICLE

REFERENCE: Meg Urry, "The Secrets of Dark Energy", The Buffalo News, Parade, Sunday, May 27, 2007

POSSIBLE TITLE OF FOLLOW-ON ARTICLE: "Buffalo New York Laboratory Identifies Dark Energy as Traveling Shock Wave Energy"

As Dr Urry notes in her above article, astronomers have been struggling, since 1998, to figure out what dark energy is.

Based on experiments I performed in the Buffalo area, I believe I have been able to figure out what dark energy is. I took a fresh approach by looking at it from an engineering (i.e. gas dynamics) perspective – a discipline where astronomers receive very little training.

I recently presented my results at a dark energy conference, at McMaster University, in Hamilton Ontario. I used the enclosed bound slides for my presentation.

I would therefore like to propose a follow-on article that identifies dark energy, based on the material in the enclosed booklet presentation.

Yours Truly

Dr. August Cenkner Jr.

<u>A Buffalo Boy:</u>

I was born and raised in the City of Buffalo. I attended Buffalo public schools and I graduated from Burgard Vocational High School. All four of my degrees were earned at the University at Buffalo, where I also taught part time in the School of Engineering. All of the referenced experiments were performed in the Buffalo area -- at the University at Buffalo and at Bell Aerospace Textron.

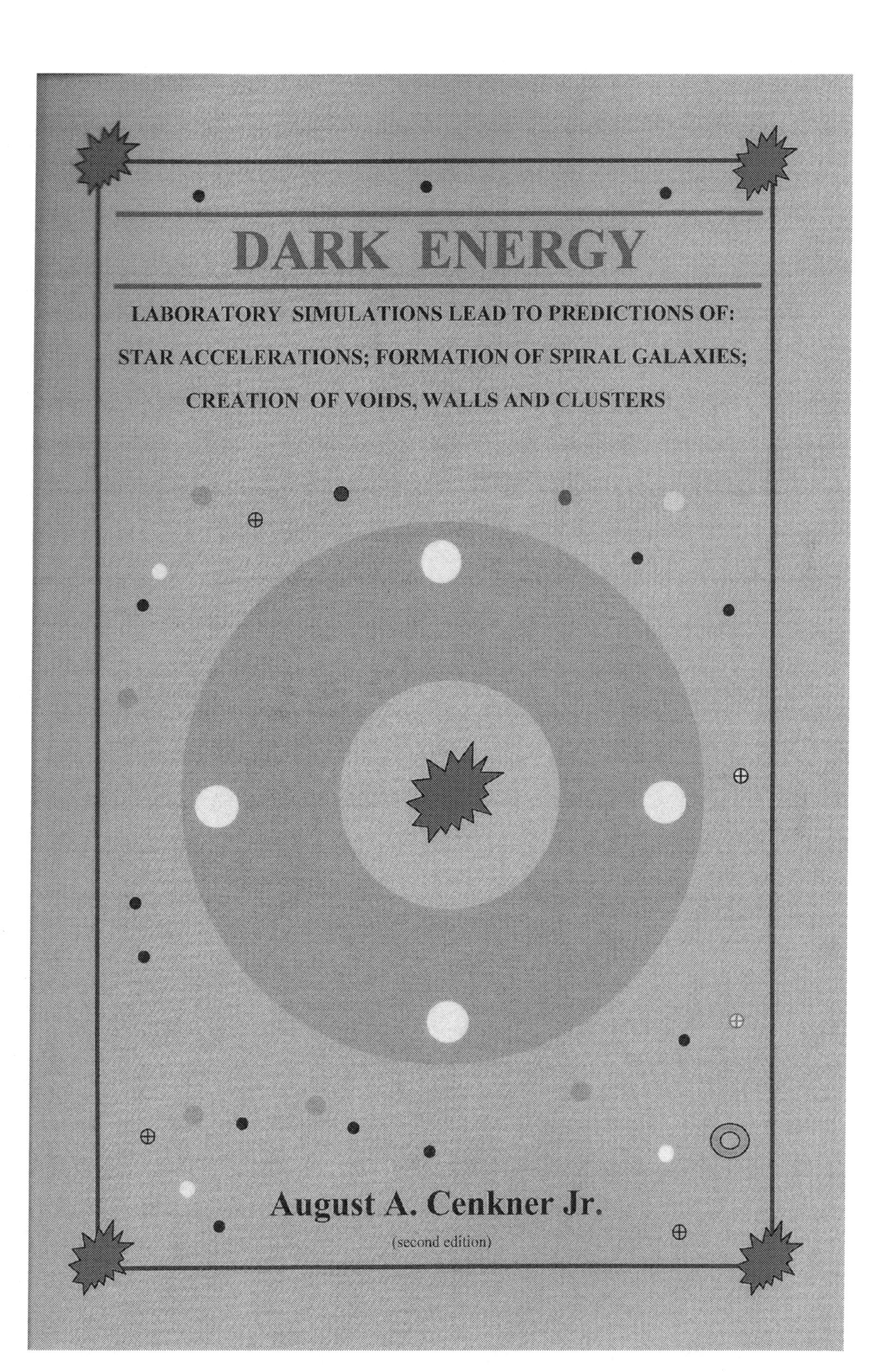
DARK ENERGY
LABORATORY SIMULATIONS LEAD TO PREDICTIONS OF:
STAR ACCELERATIONS; FORMATION OF SPIRAL GALAXIES;
CREATION OF VOIDS, WALLS AND CLUSTERS
August A. Cenkner Jr.
(second edition)

September 1, 2007

To: International Shock Wave Institute
Institute of Fluid Science
Tohoku University
Katahira, Aoba
Sendai
JAPAN

From: Dr. August Cenkner Jr.
6982 Creekview Drive
Lockport, New York 14094
USA

Subject: A Proposed New Non-traditional Shock Wave Research Area – Interdisciplinary: Gas-Dynamics/Astronomy

---

References:

(1) Cenkner, August A Jr., A Dark Energy Theory Correlated with Laboratory Simulations and Astronomical Observations, ISBN: 1420834479 (sc), AuthorHouse, 08/23/05, 94 pg.
(2) Cenkner, August A. Jr., Dark Energy – Laboratory Simulations Lead to Predictions of: Star Accelerations; Formation of Spiral Galaxies; Creation of Voids, Walls, and Clusters, ISBN: 9781434306616, AuthorHouse, Publication expected Oct 2007, 192 pg.

Background:

In 1998, astronomers determined that galaxies in the outer reaches of the universe have accelerated, instead of decelerated because of gravitational attraction with other galaxies. This acceleration is due to some mysterious unidentified force, that they have called dark energy. They are still struggling with trying to identify dark energy.

Based on my interdisciplinary background in gas-dynamics/physics/astronomy, as well as some unique experimental experience, I believe I have been able to identify dark energy. As I have detailed in my two above published books (book 1 is enclosed), I believe that dark energy is actually the energy contained in traveling shock waves.

I am led to believe that traveling shock waves have been, and continue to be, a major driving force in the evolution of the universe. For example, I think they are also involved in the establishment of other unexplained phenomena: the formation of spiral galaxies and other star groups, in addition to the formation of Voids, Walls, and Clusters.

Proposed New Shock Wave Research Area :

How Traveling Shock Waves (i.e. Dark Energy) Influence the Evolution of the Universe

I think that gas dynamics experts with experience with shock waves, and support from astronomers, should tackle this challenging, and fascinating, dark energy issue – both theoretically and experimentally. My two books (book 1 is enclosed) could provide a jumping off point. I don't think that astronomers have a strong enough background, in gas dynamics, to be able to handle this type of work on their own. Following are some areas where gas dynamic/shock wave experts could get involved.

(A) Some Possible Theoretical Studies

(1) Spherical traveling shock waves created by various types of shock emitters
  (a) Creation of shock waves by shock emitters
  (b) Impact of traveling shock waves on stars and gaseous dark bodies
  (c) Some shock emitter types
    (i) Star explosions (supernova, hypernova)
      Use various size main sequence stars, various types of gas
    (ii) Star Implosions

(iii) Star collisions

(2) 2-Dimensional traveling shock waves

(a) Shock emitter types

Jets produced by black holes

(B) Some Possible Experimental Studies

(1) Detailed study of shock impact on main sequence stars

(a) Plasma tunnels

(b) Supersonic and hypersonic shock tubes and tunnels

(2) Correlate with theoretical studies

(3) Similarity studies of star processes

(C) Astronomical Correlations

(1) Correlate predictions with information obtained by astronomers

(2) Offer suggestions to astronomers, based on theoretical predictions and laboratory simulations

Presentation:

If there is interest, I would be willing to present the contents of my books at a relevant conference. I would also be willing to distribute copies of my book at the conference.

Dark Energy

Laboratory Simulations Lead to Predictions of: Star Accelerations; Formation of Spiral Galaxies; Creation of Voids, Walls, and Clusters

This book presents a new (classical) dark energy theory. Laboratory simulations are first detailed and then used to identify dark energy as the energy contained in traveling shock waves. Finally, using Newton's Law and the First Law of Thermodynamics, it is shown how these shock waves trigger: the acceleration of stars; the evolution of spiral galaxies and other star groups; and the formation of Voids, Walls, and Clusters

August A. Cenkner Jr. B.A., B.S., M.S., Ph.D.

194 pg. ISBN 978-1-4343-0661-6 (sc) 2nd edition

Barnes & Nobel Amazon Authorhouse

* Introduced at the "Origins of Dark Energy" conference, McMaster University, Hamilton, Canada

Ad in Sky & Telescope
Feb 2008, pg 80
Vol. 115, N0.2

Ad in Sky & Telescope
July 2008, Pg 85
Vol. 116, No. 1

Dark Energy Identified

August A. Cenkner Jr. B.A., B.S., M.S., Ph.D.

Observational astronomers have concluded that galaxies, in the outer reaches of the universe, have actually accelerated. It was anticipated that these galaxies should have decelerated, due to gravitational retardation from other entities.

They have attributed this acceleration, identified in 1998, to a mysterious and unidentified repulsive force that has been labeled dark energy.

For the first time, in Reference 1, laboratory simulations were used to identify this elusive dark energy as the energy contained in traveling shock waves. These shock waves would originate from numerous violent processes, such as hypernovas, supernovas, star collisions, etc, throughout time and space.

Figure 1 illustrates the process that was to be simulated in a plasma wind tunnel. A main sequence star is gravitationally constrained by two stationary rigid dark bodies. A shock wave travels into the vacuum of space, passing by the star. The key question is "What happens to the star when the traveling shock wave passes?".

In the small-scale simulation, of Figure 2, high temperature plasma was struck between two stationary rigid electrodes. An electric field replaced the gravity field, and was used to constrain the plasma. As the shock wave passed the plasma, photographs were taken and a spectroscopic technique was used to map the isotherms in the plasma. The observed temperatures were essentially the same as the surface temperature of some main sequence stars.

It was determined that the plasma behaves like a solid deformable elastic body. There is a change in the cross sectional shape of the plasma and a low-pressure wake is formed downstream of the plasma. As the flow passes the plasma, it cannot expand fast enough to fill in this low-pressure wake. The net effect is unbalanced forces that cause the plasma to accelerate in the direction of travel of the shock.

These traveling shock waves would therefore produce star accelerations anywhere in the universe, where they occur, and not just in the outer reaches of the universe. They would act as the driving force that results in the rotational motion seen in galaxies and other star groups and in the creation of wild stars.

Horrendous violent processes, that occurred at essentially the center of the spherical-like Voids, and created powerful spherical shock waves, would eventually result in the creation of the Voids, galaxy Clusters and Walls.

Fig 1 Shock Impacts Star

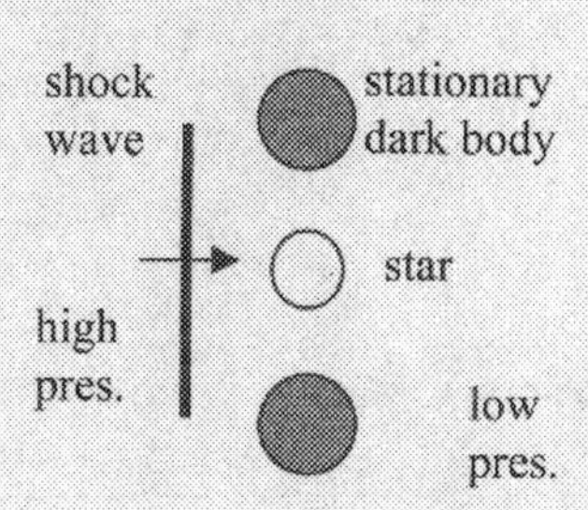

Fig 2 Laboratory Simulation

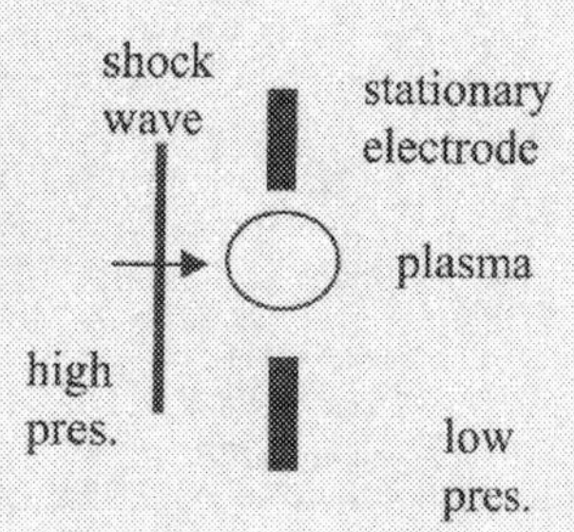

(1) Cenkner, August A. Jr, Dark Energy – Laboratory Simulations Lead to Predictions of: Star Accelerations; Creation of Spiral Galaxies; Formation of Voids, Walls, and Clusters, 2nd edition, ISBN 978-1-4343-0661-6 (sc), 194 pp, Authorhouse, 8/22/2007 (1st edition 08/23/2005)

Memorandum

To: Dark Energy Task Force

Dr. Rocky Kolb	Dr. Andreas Albrecht
Dr. Gary Bernstein	Dr. Robert Cahn
Dr. Wendy Freedman	Dr. Jacqueline Hewitt
Dr. Wayne Hu	Dr. John Huth
Dr. Marc Kamionkowski	Dr. Lloyd Knox
Dr. John Mather	Dr. Suzanne Staggs
Dr. Nicholas Suntzeff	

CC: NASA

Dr. Mary Cleave	Dr. Colleen Hartman
Dr. Paul Hertz	

Yale University

Dr. Meg Urry

Astronomy and Astrophysics Advisory Committee

Dr Neta Bahcall	Dr. John Carlstrom
Dr. Bruce Carney	Dr. Wendy Freedman
Dr. Katherine Freese	Dr. Robert Kishner
Dr. Daniel Lester	Dr. Angela Olinto
Dr. Rene Ong	Dr Sterl Phinney
Dr. Catherine Pilachowski	Dr. Saha Abhijit

TBD

From: Dr August Cenkner Jr.
6982 Creekview Drive
Lockport, NY 14094

Date: 03/25/09

Re: New (Classical) Dark Energy Theory With Laboratory Simulations

---

Dark Energy

Laboratory Simulations Lead to Predictions of: Star Accelerations; Formation of Spiral Galaxies; Creation of Voids, Walls, and Clusters

This book presents a new (classical) dark energy theory. Laboratory simulations are first detailed and then used to identify dark energy as the energy contained in traveling shock waves. Finally, using Newton's Law and the First Law of Thermodynamics, it is shown how these shock waves trigger: the acceleration of stars; the evolution of spiral galaxies and other star groups; and the formation of Voids, Walls, and Clusters

August A. Cenkner Jr. B.A., B.S., M.S., Ph.D.
194 pg. ISBN 978-1-4343-0661-6 (sc) 2nd edition
Barnes & Nobel Amazon Authorhouse

* Introduced at the "Origins of Dark Energy" conference, McMaster University Hamilton, Canada, May 2007

Unfortunately, I was unable to make your deadline for theoretical white papers. The enclosed book provides details on the laboratory simulations and the theory, as well as correlation with reported observations.

Dr. August Cenkner Jr.

The SPECTRUM Jan/Feb 2009 pg. 7
Buffalo Astronomical Association, Buffalo State College, Buffalo N.Y.

Talk for Upcoming BAA Friday Night Meeting – Date to be Announced

A Review of Astronomical Observatory Data That Relates to Dark Energy And a Demonstration of a Simulation of Dark Energy

August A. Cenkner Jr. B.A., B.S., M.S., Ph.D.

Dark energy – the unidentified repulsive force that's responsible for the acceleration of galaxies in the outer reaches of the universe – was coined after two independent teams of astronomers observed this acceleration in 1998.

A search is being conducted of the Hubble telescope database, as well as other astronomical databases, to identify any information that might be related to the identification of this dark energy. Of particular interest is any information that supports the theory that dark energy is actually the energy contained in traveling gas cloud shock waves that are associated with star explosions, etc.

This classical theory, initially proposed in Ref 1-7, is based on the results of small scale laboratory simulations in a plasma wind tunnel. The theory leads directly to a classical application of Newton's Law of Motion and the First Law of Thermodynamics, for quantitative predictions.

The search is currently concentrating on data relating to the position and velocity of stars and galaxies, and on the reported characteristics of galaxies and other star groups, including galaxy clusters and superclusters, the local flow of galaxies, walls, voids and unusual star behavior.

Spectroscopic data is relied upon to identify stars with blue-shift spectra (moving toward earth) and red-shift spectra (moving away from earth). Complicating the interpretation of blue-shift spectra is the fact that many stars are revolving stars. For example: binary, tertiary, and multiple star groups, in addition to galaxies, revolve around their center of mass. Furthermore galaxies, in galactic clusters, revolve around their cluster center of mass. To challenge us even further, we also have to throw in the Hubble recessional velocity, which is the velocity associated with the expansion of the universe, and earth's motion.

Current findings will be discussed, along with the implications of how these traveling shock waves can affect the future of the universe.

Finally, a time line will be suggested to show how traveling shock waves could have influenced the evolution of the universe, throughout space and time, starting with the "Big Bang" or "Rapid Expansion".

References

(1) Cenkner, August A. Jr., "Dark Energy Identified", Sky and Telescope, 7/8/08, pg. 85.
(2) Cenkner, August A. Jr, "Dark Energy – Laboratory Simulations Lead to Predictions of: Star Accelerations; Creation of Spiral Galaxies; Formation of Voids, Walls, and Clusters", 2nd edition, ISBN 978-1-4343-0661-6 (sc), 194 pgs, Authorhouse, 8/22/07.
(3) Cenkner, August A. Jr., "A Dark Energy Theory", Origins of Dark Energy Conference, McMaster University, Hamilton, Ontario, Canada, 6/14/07.
(4) Cenkner, August A. Jr., "A Dark Energy Theory Correlated With Laboratory Simulations And Astronomical Observations", ISBN 1-4208-3447-9 (sc), 98 pgs. Authorhouse, 8/23/05.
(5) Cenkner, August A. Jr., "Galaxies", Buffalo Astronomical Association, Buffalo State College, Buffalo, NY, 6/12/05.
(6) Cenkner, August A. Jr., "Universe Acceleration Predictions Using New Dark Energy Explanation", Buffalo Astronomical Association, Buffalo State College, Buffalo NY, 7/10/04.
(7) Cenkner, August A. Jr., "A Theory to Explain Dark Energy", The Spectrum, Buffalo Astronomical Association, Buffalo State College, Buffalo NY, 3/7/04.

The SPECTRUM May/June 2009 pg. 1 of 4
Buffalo Astronomical Association, Buffalo State College, Buffalo N.Y.

HUBBLE SPACE TELESCOPE SEES DARK ENERGY*

August Cenkner Jr. B.A., B.S., M.S., Ph.D.

I am searching the Hubble Telescope database for any information that might reveal insight into the origins of dark energy. Of particular interest is any information that might be related to the new classical dark energy theory (Ref 1-8) that has identified dark energy as the energy contained in traveling gas cloud shock waves, which originated from violent processes that eject gas clouds, like star explosions, star collisions, etc. Laboratory experiments have revealed that the stars are being pushed by this high pressure traveling gas because a low pressure wake is formed on the downstream side of the star. Of particular interest, to this study, are unusual star motions that are currently unexplained, in addition to any regions of space that are free of galaxies and/or stars.

As we will see, star pushing by the traveling gas cloud shock wave, offers a plausible explanation for a number of observed but unexplained phenomena.

(1) Hoag's Ring Galaxy (Hubble STSci-PRC-2002-21) a circular ring galaxy with an outer ring of stars (that are moving outward), an inner ring free of stars, and a core region with a group of older stars – with no nearby galaxy that could have collided with it and no current explanation on how it was created; see Fig. 1.

TRAVELING-SHOCK-WAVE-PUSHING EXPLANATION: Using the dark energy traveling-shock-wave-pressure theory, a violent explosion occurred at the center of a disc galaxy, creating a spherical traveling radial gas cloud shock wave whose pressure accelerated stars in the disc and pushed them outward radially, creating an inner disc that was free of stars. The remnant of this star explosion (e.g. a black hole or a dark body) then attracted older passing stars that began orbiting this core remnant. This explanation may apply to other ring galaxies.

This type of process would have also resulted in the creation of the Voids, as shown in Fig. 4.

(2) Colliding Galaxies in Great Wall of Clusters and Superclusters: NGC6050 & IC 1179 Spiral Galaxies (Hubble Heio0810ap) see Fig. 2.

TRAVELING-SHOCK-WAVE-PUSHING EXPLANATION: In Ref 2&4 it is shown how violent explosions, occurring at the center of the Voids, would have created gas cloud shock waves that pushed all of the stars out of the void regions, in a radial direction. Since there are voids on at least two sides of each Wall, some galaxies from different Void regions would therefore be traveling toward one another. It is reasonable to expect that some of these galaxies would collide and some would begin orbiting each other.

(3) M13 Globular Cluster (Hubble 12-2-08) – a group of several hundred thousand stars orbiting in various directions around a common center of mass; see Fig. 3.

TRAVELING-SHOCK-WAVE-PUSHING-EXPLANATION: A violent gas ejecting process created a traveling shock wave that accelerated millions of assorted sized stars in various directions. In a given direction, the faster stars overtook the slower stars (or objects) and were gravitationally captured into an orbit around the slower object. Thousands of stars were captured this way, creating randomly oriented orbits and culminating in what is known as the M13 globular cluster.

A similar process would have occurred for all star groups and galaxies having revolving stars; see Ref. 2 & 4.

It is also worth pointing out several areas where the traveling-shock-wave-pushing-theory offers an alternate, but plausible, explanation. Addition study would be required to determine if shock-wave-pushing, gravity, or galaxy collisions (which may result from shock-wave-pushing) dominate.

(8) ABOUT THE AUTHOR

Dr. August A. Cenkner Jr. has initiated a technology transfer activity, where he is transferring known gas dynamic technology to the astronomical area -- to shed light on unexplained phenomena in the astronomical area and to point the direction for the interpretation of astronomical observatory data..

He received extensive graduate education in gas dynamics and he has extensive theoretical and experimental experience in gas dynamics. He earned a B.A. degree in Computer Science, and a B.S. degree in Aerospace Engineering, along with M.S. and Ph.D. degrees in Engineering Science. His graduate studies were concentrated in the area of advanced gas dynamics. It encompassed: Supersonic flow (1&2), Hypersonic Flow (1&2), Non-Steady Gas Dynamics, Rarefied Gas Dynamics, Plasma Physics, Magnetohydrodynamics, Aerothermochemistry, Radiation Heat Transfer, Radiation Gas Dynamics, Electrodynamics, Direct Energy Conversion, Diagnostic Techniques, and Space Science. Non-credit courses were also attended: Spectroscopy, Optics, Astronomy (1&2), and Geology.

For forty years, Dr. Cenkner was involved in research and development -- performing theoretical, experimental, and design work. During this period, he acquired laboratory diagnostic and interpretation skills while working on 2 plasma wind tunnels, 2 supersonic wind tunnels, a shock tube, a gas dynamic laser, compressors, vacuum chambers and a subsonic climatic wind tunnel. As an adjunct faculty member, he taught 32 courses in the school of engineering, including: statics, dynamics, thermodynamics, fluid mechanics and heat transfer. He has twenty one publications, covering research on gas dynamic systems, in the open literature.

Previously, he has published two earlier editions of his dark energy book, detailing progress on his dark energy project.

Over the years, Dr. Cenkner has also studied astronomy, astrophysics, and cosmology. He is a long time amateur astronomer who is actively involved in the Buffalo Astronomical Association.

He is a firm believer in technology transfer because nature has not conveniently compartmentalized itself into human defined compartments.

Dr. August A. Jr. and Mrs. Judith B. Cenkner